Supported by Projects of Bilingual Teaching and
Key Textbook Construction for Undergraduate Students (CAU)

Introduction to Bioenvironmental Engineering

Built Environment

■ Editor **Yuanhui Zhang**

■ Associate Editors **Wei Cao**
Chaoyuan Wang
Doug Barker

CHINA AGRICULTURAL UNIVERSITY PRESS
· BEIJING ·

图书在版编目(CIP)数据

生物环境工程导论:建筑环境 = Introduction to Bioenvironmental Engineering-Built Environment:英文/张源辉主编.—北京:中国农业大学出版社,2010.12

ISBN 978-7-5655-0140-1

Ⅰ.①生… Ⅱ.①张… Ⅲ.①建筑工程-环境管理-教材-英文 Ⅳ.①TU-023

中国版本图书馆 CIP 数据核字(2010)第 231391 号

书　名 Introduction to Bioenvironmental Engineering-Built Environment

作　者 Yuanhui Zhang 主编　Wei Cao, Chaoyuan Wang, Doug Barker 副主编

策划编辑 童　云　　**责任编辑** 田树君

封面设计 郑　川

出版发行 中国农业大学出版社

社　址 北京市海淀区圆明园西路 2 号　　**邮政编码** 100193

电　话 发行部 010-62731190,2620　　读者服务部 010-62732336

编辑部 010-62732617,2618　　出　版　部 010-62733440

网　址 http://www.cau.edu.cn/caup　　**E-mail** cbsszs@cau.edu.cn

经　销 新华书店

印　刷 北京时代华都印刷有限公司

版　次 2010 年 12 月第 1 版　2010 年 12 月第 1 次印刷

规　格 787×980　16 开本　13.5 印张　240 千字

定　价 28.00 元

Editor: Yuanhui Zhang (China Agricultural University, China;
University of Illinois, USA)

Associate Editors: Wei Cao (China Agricultural University, China)
Chaoyuan Wang (China Agricultural University, China)
Doug Barker (University of Illinois, USA)

2010

This book is financially supported by the Project of Bilingual Teaching and Key Textbook Construction for Undergraduate Students, China Agricultural University.

Preface

Built environment affects many aspects of our lives-quality of living, energy consumption and subsequent carbon emission to the atmosphere. In the United States, building sector, including residential and commercial, is the largest energy consumer using 40% of the total energy compared to 29% for transportation sector and 31% for industrial sector in the past decade. Built environment principles are indispensable elements in the body of knowledge of engineering students who are interested in areas of building, energy conservation, and indoor environment quality.

This book is an accumulation of the lecture notes during the past 16 years of teaching as an undergraduate introduction course at the University of Illinois. The objectives are, by completing this course, to enable the students to: (1) use a psychrometric chart, measure and calculate psychrometric properties such as temperatures, relative humidity, enthalpy, sensible heat and latent heat; (2) determine steady-state heat transfer rates in three modes: conduction, convection and radiation, through building envelopes, and calculate the heating/cooling loads; and (3) determine the ventilation requirements for the control of temperature, humidity and airborne pollutants, and select heating/ventilation equipment. This course has also been taught as a duel-language course at the College of Water Conservancy & Civil Engineering at China Agricultural University. Although this book is primarily written for engineering students, the principles and approach is suitable for students in related disciplines who are interested in building environment control and energy conservation.

As a textbook, it is suggested to be taught as a 2 to 3 hour credits course at sophomore or junior level. The twelve modules can be divided into three sections: (1) psychrometrics; (2) building heat transfer basics; and (3) mass transfer and ventilation requirement. First section includes Modules 1 to 4 that describes psychrometric chart for achieving Objective 1 of the course. Second section includes Modules 5, 6 and 7 that introduces heat transfer so the students can calculate the heating and cooling loads through a building envelope (Objective 2). The third section includes Modules 8 to 12 that discusses the ventilation requirement for an airspace to maintains energy and mass balances (Objective 3). There are homework assignments and laboratory exercises for each section. Several case studies in Module 12 are designed for classroom discussions or team studies.

Many graduate students have contributed to the lecture notes. I am in particular to acknowledge the co-authors of this book, Doug Barker, University of Illinois, Drs. Wei Cao and Chaoyuan Wang, China Agricultural University, for their comments, contributions to working problems and case studies.

Yuanhui Zhang
Urbana, IL
October 22, 2010

Contents

Module 1: Introduction

Air is a critical element to many living things. People, for example, may survive for two weeks without food and for two days without water. But without air, a person may only survive for two minutes! In daily life, an average person consumes approximately one kilogram of food, two kilograms of water and 20 kilograms of air! Furthermore, in developed countries, an average working person spends over 90% of their lifetime indoors.

Table 1-1. Daily intake of air of an average person and hog

Average Daily Intake		

	1 kg(2.2 lb)	2 kg(4.4 lb)
	2.0 kg(4.4 lb)	4 kg(8.8 lb)
	20 kg(44 lb)	40 kg(88 lb)

Most people are aware that outdoor air pollution can damage the environment and their health, but may not know that indoor air pollution can also have significant effects. Until the late 1960s, attention to air quality was primarily focused on outdoors because at that time, outdoor air pollution was considered to be responsible for many adverse health problems. In the early 1970s, scientists started to investigate the cause of complaints of indoor working environments. U. S. EPA studies of human exposure to air pollutants indicated that indoor air levels of many pollutants may be 2-5 times, and occasionally more than 100 times, higher than outdoor levels. Over the past several decades, our exposure to indoor air pollutants is believed to have increased due to a variety of factors, including the construction of more airtight buildings, reduced ventilation rates to conserve energy, the use of synthetic building materials and furnishings, and the use of chemically formulated personal care products, pesticides, printing inks, and household cleaners. In recent years, comparative risk studies performed by the U. S. EPA and its Science Advisory Board (SAB) have consistently ranked indoor air pollution among the top five environmental risks to public health. As a result of these studies, people began to realize that indoor air quality is important to the comfort and health of people. Additionally, improvement in the quality of outdoor air since the 1972 Clean Air Act enhances the relative importance of indoor air quality.

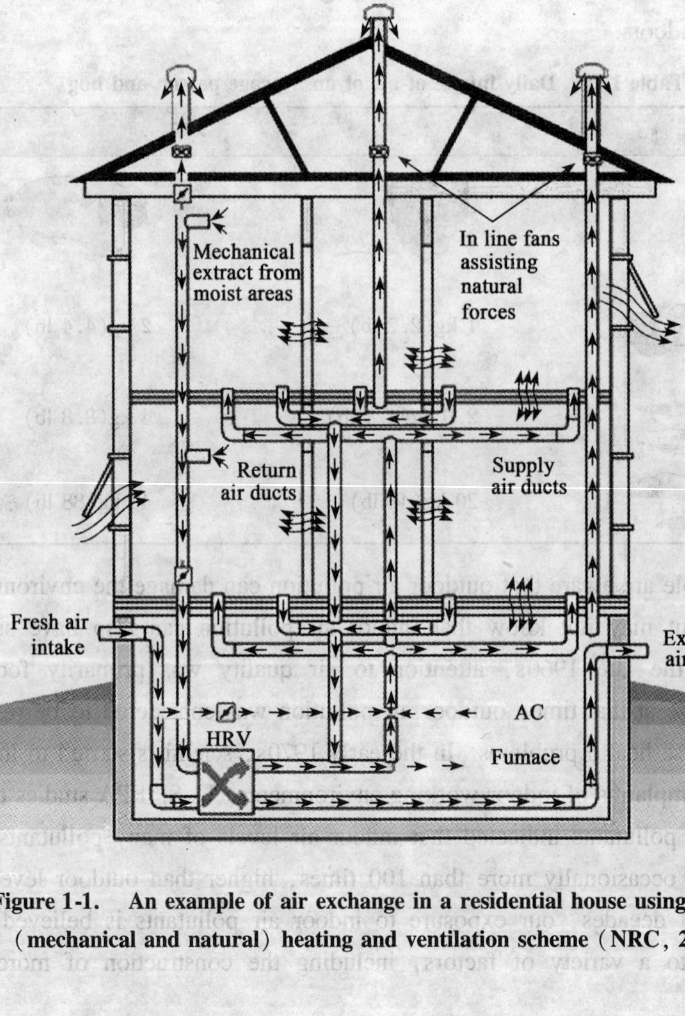

Figure 1-1. An example of air exchange in a residential house using hybrid (mechanical and natural) heating and ventilation scheme (NRC, 2006)

The importance of indoor air quality is also due to the sheer amount of time that people spend indoors. Traditionally, people spent more time outdoors than indoors. Today, people spend more than 90% of their lifetime indoors in industrialized countries. Take the United States for example, every day an average working person spends 22 hours and 15 minutes indoors and one hour in cars or other types of transportation mode-another type of indoor environment. Thus, the total time spent indoors for an average working person is 23 hours and 15 minutes per day, or 97.7% of his/her lifetime! Other types of people such as young children and seniors may spend more time indoors than employed individuals. Indeed, the quality of indoor air should be as high as possible.

1.1 Air Composition

In order to define air quality, a baseline reference to "clean air" or "standard air" should first be established. A typical clean air is the dry atmosphere air found in rural areas or over the ocean far from air pollution sources. The chemical composition of such clean dry atmospheric air is listed in Table 1-2. The atmospheric air also contains from 0.1% to 3% of water vapor by volume depending on temperature. The clean air defined in Table 1-2 is typical because in many instances, other traces of components are also found in the atmosphere that is considered clean. These other tracer components include ammonia, sulfur dioxide, formaldehyde, carbon monoxide, iodine, sodium chloride, and particulate matters such as dust and pollens.

Based on the definition of clean air, air quality refers to the degree of pollution of the clean air. In general, the lower the concentration of airborne pollutants, the better the air quality. *Airborne pollutant* is defined as the substances in the air that can harm the health and comfort of humans and animals, reduce performance and production of plants, or accelerate damage to equipment. Airborne pollutants can be in the form of solid (e. g., particulate matters), liquid (e. g., mists) and gaseous substances. Excessive high concentrations or depletion of substances listed in Table 1-2 can impose serious air quality problems. For example, it is suspected that excessive emissions of carbon dioxide and methane are responsible for greenhouse effects and global warming.

Table 1-2. Chemical composition and volumetric content of typical dry atmospheric air

Substances	Content (%[a])	Concentration (mg/L[a])
Nitrogen	78.084 ±0.004	780 840
Oxygen	20.946 ±0.002	209 460
Argon	0.934 ±0.001	9 340
Carbon dioxide	0.033 ±0.001	330
Neon		18
Helium		5.2
Methane		1.2
Krypton		0.5
Hydrogen		0.5
Xenon		0.08
Nitrogen dioxide		0.02
Ozone		0.01-0.04

Source: *The Handbook of Air Pollution*, PHS Publication, 1968.

[a] in volume.

1.2 Variables and Units

Throughout the text, the SI (International Systems) units-meter, kilogram and second (m-kg-s), will be used as the primary dimension system. However, due to the nature of particle size which is usually in the range of microns or sub-microns, centimeter, gram and second (c-g-s) are used as a secondary dimension system. Liters and milliliters are also used in very small volume or low flow rate. Those secondary units are easily derived from the m-kg-s system. It is important to pay special attention to the consistency of units before applying an equation.

1.3 Energy Use for Buildings

Heating, ventilation and air-conditioning constitute a large portion of the national energy consumption. The following summary data are from the U. S. Department of Energy Data Book (DOE, 2008). From Figure 1-2, energy used in buildings-residential and commercial combined-represents 39.84% of the total energy of the country. It is approximately the total petroleum energy (40.44% of total).

The United States consumed 1.2 billion tons of crude oil (57% imported) in 2007. To put it in perspective, on average, building energy used per person per year in the United States topped approximately four tons of crude oil equivalent. The U. S. Department of Energy summarizes and predicts building energy consumption annually at the website: http://buildingsdatabook.eren.doe.gov. Additional data on building energy use published in 2008 is in Appendix 1.

Table 1-3. Psychrometric variables and their symbols and dimension

Definition	Symbol	Dimension
Temperature:		
Dry bulb temperature	T	K or ℃
Wet bulb temperature	T_w	K or ℃
Dew point temperature	T_d	K or ℃
Humidity:		
Humidity ratio	W	kg_w/kg_a
Relative humidity	Φ	%
Degree of saturation	μ	%
Enthalpy:	h	kJ/kg
Sensible heat content	h_a	kJ/kg
Latent heat content	h_w	kJ/kg
Pressure:	P	kPa
Water vapor saturation partial pressure	P_{ws}	kPa
Water vapor partial pressure	P_w	kPa
Atmosphere pressure	P	kPa
Heat transfer:		
Heat transfer rate	q	J/s or W
Heat flux	q''	W/m^2
Mass transfer:		
Mass	M	kg
Mass flow rate	m	kg/s
Volumetric flow rate	Q	m^3/s
Specific volume	v	m^3/kg
Density	ρ	kg/m^3

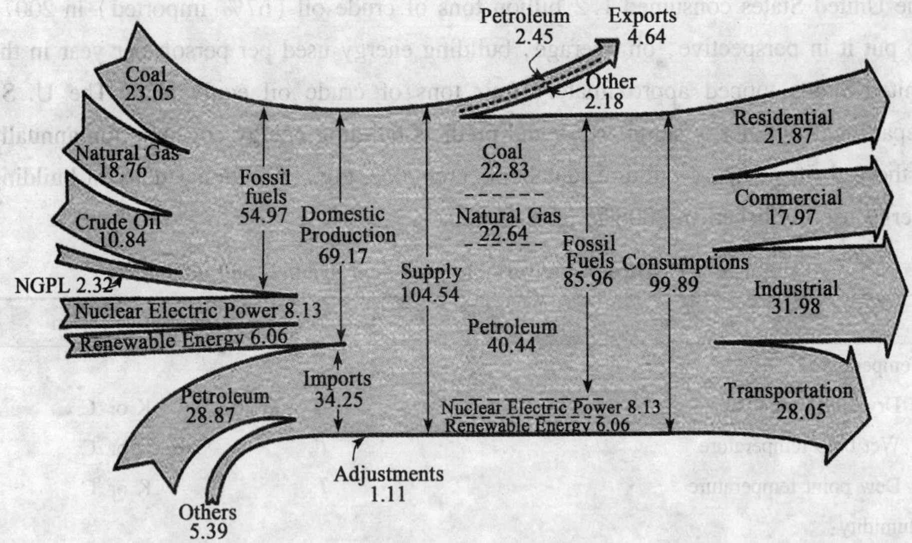

Figure 1-2. Energy flow through the United States in 2005 (EIA, 2005)

1.4 Course Objectives

By completing this course, you will have a basic understanding of psychrometric, heat transfer through a building shelter and ventilation. Specifically, you will be able to:

- Use a psychrometric chart, measure and calculate psychrometric properties, including temperature, relative humidity, enthalpy, sensible heat, and latent heat.
- Determine steady-state heat transfer rates in three modes: conduction, convection and radiation, through different building shelters.
- Calculate ventilation requirements for control of temperature, humidity and airborne contaminants, and properly size heating/ventilation equipment.

1.5 Laboratory Report Policies

Effective technical writing is an essential skill for engineers in all fields. Engineers who can communicate their work clearly, concisely and accurately to a wide range of audiences will be more successful in the long run.

"Even the most well designed and flawlessly executed technical project is not complete or significant until the results are communicated effectively."

Format: Your lab reports will be formatted similar to common company project reports, engineering journal papers and proposals to government agencies. The following page gives a general list of contents required for each lab report. There is a ten-page limit (excluding the cover page and appendices).

Grading: Reports are due 1 week from the completion of the Lab. The reports will be graded according to the accuracy of the technical content, level of effort and the student's ability to clearly and concisely convey technical information. This includes spelling, grammar, composition, and overall flow.

Teamwork Policy: In the real world you will rarely work alone on a project. Therefore, teamwork is strongly encouraged during the lab for collecting and analyzing data. However, lab reports must be written individually; therefore, you must be able to explain what your team did and the results in your own words. You must reference any sources you used in the lab report. Copying results in zero credit.

1.6 Recommended Laboratory Report Format

1. **Cover Page**: Title (Lab title and #), course name and number, your name and department, other team members, and date.

2. **Introduction (10 pts.)**: Explain the context of the information you are about to present, a brief background, why it is important, etc.

3. **Purpose and Objectives (10 pts.)**: Explain the purpose of the work in the report and list the key objectives. *What is this work trying to accomplish*?

4. **Methods and Approach (30 pts.)**: Here is where you describe the approach you used to achieve your listed objectives. *What you did and how*? Include: equipment and/or instrumentation used, procedures, experimental design, analytical methods (key equations and assumptions to be used) and any additional tactics used to achieve your objectives.

5. **Results and Discussion (40 pts.)**: Report any results or findings yielded from the lab. Summarize the calculations and data. Interpret the data and explain

what it means. If you include a graph, interpret what it is showing and what may have contributed to this result. Answer the question: *Is this result reasonable*? Explain why or why not. *What are potential sources of error*? Be sure to keep your objectives in mind when writing this section. *Have you achieved the objectives*?

6. **Conclusions (10 pts.)**: Briefly summarize your major findings or results in this section and clearly state the conclusions that can be made.

7. **References**: List any books, papers, web pages (with date accessed), etc., that you obtained information from.

Appendix: Long derivations, raw data, non-essential figures (excess graphs/tables) should be included here. For example if you have a table that takes up over a page, it should probably be in the appendix and not the body of the report. Please note that all information in the appendix must be cited in the body to inform the reader of its existence and why its there.

1.7 Lab Report Example

Lab 1: Psychrometrics

By

Date

1.7.1 Introduction

Psychrometrics plays a vital role in our every day from maintaining human comfort to the performance of engineering systems. In this lab, we try to gain a better understanding of the methods for measuring and calculating the psychrometric properties of air.

1.7.2 Purpose and Objectives

The purpose of the lab is to measure the dry-bulb and wet-bulb air temperature and use these data to determine other psychrometric properties. The objectives were to:

1. Learn how to use the psychrometer and measure the room conditions.
2. Calculate the properties in the room under a given set of conditions.
3. Solve for a new room condition after 10 minutes when students are present.

1.7.3 Methods and Approach

A psychrometer is used to measure temperature:

- The *dry-bulb temperature* is read directly from a thermometer.
- The *wet-bulb temperature* is read some minutes after wetting the thermometer bulb end and engaging a blower to evaporate the water.

Other psychrometric properties are determined from the above measured values and a psychrometric chart:

- Point A is located at the convergence of two lines normal to the wet-and dry-bulb temperature axes and passing through the measured wet-and dry-bulb temperatures, respectively.
- The *relative humidity* may be interpolated from the relative location of point A and the known relative humidity curves on the chart.
- The *humidity ratio* may be interpolated from the relative location of point A and the known humidity ratio lines on the chart and converting the observed value $\left(\frac{g_{H_2O}}{kg_{dry\_air}}\right)$ to a true ratio $\left(\frac{kg_{H_2O}}{kg_{dry\_air}}\right)$ by dividing by 1 000.
- The *enthalpy* may be interpolated from the relative location of point A and the known enthalpy lines on the chart.
- The *dew-point temperature* may interpolated from the relative location of point B and the known wet-bulb temperature lines on the chart; point B is located at the intersection of the wet-bulb temperature axis and a horizontal line passing through point A.

- The *sensible heat* is the enthalpy at point C, located at the dry-bulb temperature and 0% relative humidity.
- The *latent heat* may be calculated by taking the difference between the enthalpy and the sensible heat.

1.7.4 Results and Discussion

The following results were obtained from the psychrometer readings and chart. All calculations used are shown in the appendix.

Table 1-4. Psychrometric properties of room air during the lab session

Parameter	Symbol	Value	Units
Dry bulb temperature	T_d	26	℃
Wet bulb temperature	T_w	15	℃
Dew point	$T_{dewpoint}$	6.5	℃
Relative humidity	Φ	30	%
Humidity ratio	W	0.006 3	g H_2O/kg dry air
Total enthalpy	h	42	kJ/kg
Sensible enthalpy	h_s	26	kJ/kg
Latent enthalpy	h_w	16	kJ/kg

The results for the second part of the lab are shown below. For this section a room, which is sealed and fully insulated, that contains 100 kg moist air and 15 people was analyzed. Each individual generated 100 W total heat and 40 g/hr of moisture. After 10 minutes the condition of the room air was identified and it was determined if a problem exists in the room, and a potential solution was offered in the section below.

The following parameters were calculated:

- Sensible enthalpy added into the room
- Latent enthalpy added into the room
- Temperature in room
- Relative humidity in room

Additional moisture:

$$15\ \text{person} \times \frac{40\ \frac{g_{H_2O}}{h}}{\text{person}} \times \frac{h}{60\ \text{min}} \times 10\ \text{min} \times \frac{\text{kg}}{1\ 000\ \text{g}} = 0.1\ \text{kg}_{H_2O}$$

Additional heat:

$$15\ \text{person} \times \frac{100\ \text{W}}{\text{person}} \times \frac{\text{kW}}{1\ 000\ \text{W}} \times \frac{3\ 600\ \dfrac{\text{kJ}}{\text{h}}}{\text{kW}} \times \frac{\text{h}}{60\ \text{min}} \times 10\ \text{min} = 900\ \text{kJ}$$

Mass dry air:

$$100\ \text{kg}_{\text{wet\_air}} = Wx + x$$
$$x = 99.4\ \text{kg}_{\text{dry\_air}}$$

Additional heat per unit mass:

$$h_{\text{add}} = \frac{900\ \text{kJ}}{99.4\ \text{kg}_{\text{dry\_air}}} = 9.1\ \frac{\text{kJ}}{\text{kg}_{\text{dry\_air}}}$$

Total heat per unit mass:

$$h' = h + h_{\text{add}} = (42 + 9.1)\frac{\text{kJ}}{\text{kg}} = 51\ \frac{\text{kJ}}{\text{kg}}$$

New humidity ratio:

$$W' = W + \frac{0.1\ \text{kg}_{\text{H}_2\text{O}}}{99.4\ \text{kg}_{\text{dry\_air}}} = 0.007\ 3$$

Combining h' and W' to find point A (on psychrometric chart), we get:

$$T' = 32.2℃$$
$$\Phi' = 24\%$$
$$h_a{}' = 33\ \frac{\text{kJ}}{\text{kg}}$$

and then:

$$h_{\text{w}}{}' = (51.1 - 33)\frac{\text{kJ}}{\text{kg}} = 18\ \frac{\text{kJ}}{\text{kg}}$$

Thus, to get the added sensible and latent enthalpies:

$$h_{\text{a\_add}} = h_{\text{a}}{}' - h_{\text{a}} = 33 - 26 = 7\ \frac{\text{kJ}}{\text{kg}}$$

$$h_{\text{w\_add}} = h_{\text{w}}{}' - h_{\text{w}} = 18 - 16 = 2\ \frac{\text{kJ}}{\text{kg}}$$

The answers to the question are as follows.

$$h_{\text{a\_add}} = 7\ \frac{\text{kJ}}{\text{kg}}$$

$$h_{\text{w\_add}} = 2\ \frac{\text{kJ}}{\text{kg}}$$

$$T' = 32.2℃$$

$$\Phi' = 24\%$$

1.7.5 Discussion questions

Is there any problem in the room after 15 *minutes*?

• The temperature (about 90°F) may well be considered excessive by many according to the ASHRAE standard comfort zones for winter. Air conditioning would provide cooler air, and, considering the season (March), opening the windows and doors would introduce cooler outside air.

If there is, offer your solutions to solve the problem. (Depending on the nature of the problem in the room, solutions may include heating, cooling, adiabatic mixing, and air cleaning).

• The final relative humidity is actually less than the initial relative humidity due to the increased water-holding capacity of the warmer air. Somehow cooling this air without removing any moisture may result in condensation and would likely result in uncomfortably "muggy" air. However, ventilating the room through the windows would replace the "wet" air with air at the outside moisture content; in March, the relative humidity is generally not onerous. Air conditioning would remove some water by cooling a portion of the air to a low enough temperature to cause condensation in the air conditioner, removing that water from the room's air. Of course, should this room be on a space station, or submarine, or some other such sealed environment, a strongly hygroscopic material could be used to absorb water from the air.

1.7.6 Conclusions

• Temperature values from the two psychrometers varied slightly but were generally close to each other. Additionally, wet-bulb temperature readings varied over time, as the water in the psychrometer evaporated.

• The dew-point temperature is well below the wet-and dry-bulb temperatures, as expected for a low relative humidity.

• Values were generally limited to two significant digits; interpolating these data from the chart with more significant digits would be erroneously precise, especially as the initial temperature readings varied slightly. Should more significant digits be desired, a computer program should be used.

• Values from the chart were given for sea level atmospheric pressure, though central Illinois is above sea level. In actual engineering practice, these values would be corrected for the pressure difference.

1.7.7 Appendix: Calculations

- Temperature conversions

$$T_C = \frac{T_F - 32}{1.8}$$

$$T = \frac{79 - 32}{1.8} = 26℃$$

$$T_w = \frac{59 - 32}{1.8} = 15℃$$

- Humidity ratio:

$$W = \frac{6.3\ \frac{g_{H_2O}}{kg_{dry\_air}}}{1\ 000} = 0.006\ 3\ \frac{kg_{H_2O}}{kg_{dry\_air}} = 0.006\ 3$$

- Latent enthalpy:

$$h_w = h - h_a = (42 - 26)\frac{kJ}{kg} = 16\ \frac{kJ}{kg}$$

1.8 Homework Format

Known: This is a paraphrased statement of the problem given in the homework. Try to distill the problem into only the vital information needed to solve the problem and restate that here.

Find: Write down what exactly the problem wants you to find. Use a numbered list if needed.

Given: Write down all of the data given in the problem statement to provide a clear understanding of the known values that you have to work with. Often this will be system parameters such as temperature and pressure or material properties such as density or specific heat.

Solution:

- System diagram (if applicable)
- Governing equation
- Assumptions
- Known's and unknowns

- All equations and calculations used to solve for the unknown values
- Boxed answer with appropriate units

1.9 Homework Example

Known: A wall is made of three layers. The outer layer is 10 cm of concrete block (rectangular core, sand and gravel aggregate, 2 cores), the inner layer is the same and the center layer is made of 5 cm of expanded polystyrene molded beads with a density of 16 kg/m^3.

Find: Total thermal resistance of the wall, *RSI*.

Given:

Layer # 1-Outer layer	Layer # 2-Center layer	Layer # 3-Inner layer
$\Delta x_1 = 0.1$ m	$\Delta x_2 = 0.05$ m	$\Delta x_3 = 0.1$ m
Concrete block	Expanded polystyrene beads	Concrete block
	$\rho = 16$ kg/m^3	

Solution: System diagram

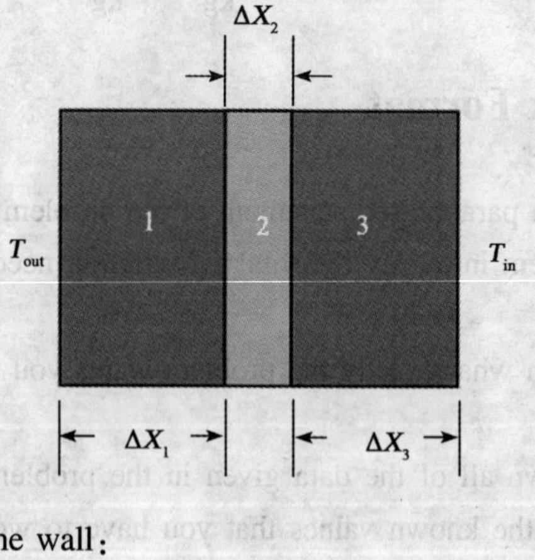

Find total *RSI* for the wall:

Use Equation 1-1 for a thermal resistance of a multi layer wall.

$$RSI_{\text{total}} = R_1 + R_2 + R_3 \tag{1-1}$$

Unknowns: R_1, R_2, R_3

Calculate unknowns using information from Appendix 3:

$$R_1 = \frac{RSI_{\text{rated}}}{\Delta x_{\text{rated}}} \cdot \Delta x_1 = \frac{0.18}{203.2\ \text{mm}} \cdot 100\ \text{mm} = 0.088\,57\ \text{m}^2 \cdot ℃/\text{W}$$

$$R_2 = \frac{RSI_{\text{rated}}}{\Delta x_{\text{rated}}} \cdot \Delta x_2 = \frac{23.25}{1\ \text{m}} \cdot 0.05\ \text{m} = 1.162\ 5\ \text{m}^2 \cdot ℃/\text{W}$$

$$R_3 = R_1 = 0.088\ 57\ \text{m}^2 \cdot ℃/\text{W}$$

$$RSI_{\text{total}} = 0.088\ 57 + 1.162\ 5 + 0.088\ 57 = 1.339\ \text{m}^2 \cdot ℃/\text{W}$$

Module 2: Psychrometrics

Psychrometrics or psychrometry are terms used to describe the field of engineering concerned with the determination of physical and thermodynamic properties of gas-vapor mixtures. The term derives from the Greek *psuchron* (ψυχρόν) (cold) and *metron* (μέτρον) (means of measurement). The principles of psychrometry apply to any physical system consisting of gas-vapor mixtures, including air. The most common systems of interest, however, are mixtures of water vapor and air because of its application in heating, ventilation and air-conditioning and meteorology. The application of psychrometrics is also found in grain drying, food processing and the design of many biological systems and equipment. It is one of the most basic and widely used disciplines that one can relate to in his/her daily life.

2.1 Psychrometric Properties

Psychrometric properties of moist air refer to temperature, humidity, heat content, enthalpy, and specific volume. Of all the properties, temperature and relative humidity are most commonly used. All psychrometric properties can be obtained from any two known properties, such as temperature and relative humidity. Temperature can be measured using a thermometer; relative humidity can be measured using a psychrometer.

A psychrometric chart is a graph of the physical properties of moist air at a constant pressure (often equated to an elevation relative to sea level). The chart graphically expresses how various properties relate to each other, and is thus, a graphical equilibrium of state. The thermophysical properties found on most psychrometric charts, also called psychrometric properties, are described in Table 2-1.

A psychrometer is an instrument for measuring the relative humidity of air (Figure 2-1). It consists of two thermometers: a dry-bulb thermometer and a wet-bulb thermometer. The difference between the readings on the thermometers is used to determine the relative humidity. Psychrometric readings and their interpretation are important in determining proper ventilation.

Table 2-1. Definition of psychrometric properties

Dry-bulb temperature	Temperature is measured by a thermostat. It is an indicator of sensible heat in dry air. (When air is not specified, it refers to moist air = mixture of dry air and water vapor.)
Wet-bulb temperature	Temperature is obtained through evaporation adiabatically of moisture from the wick on a thermometer
Dew-point temperature	A temperature at which condensation is just starting to occur
Relative humidity	The ratio of actual water vapor pressure to saturated water vapor pressure at the same temperature
Humidity ratio (absolute or specific humidity)	The ratio of the mass of water vapor to the mass of dry air
Degree of saturation	The ratio of actual mass of water vapor to the mass of saturated water vapor at the same temperature
Enthalpy	The total heat contained in the air
Sensible heat	The heat contained in dry air
Latent heat	The heat contained in water vapor
Specific volume	The volume of air per mass of air
Density	The mass of air per volume of air (or the reciprocal of specific volume)

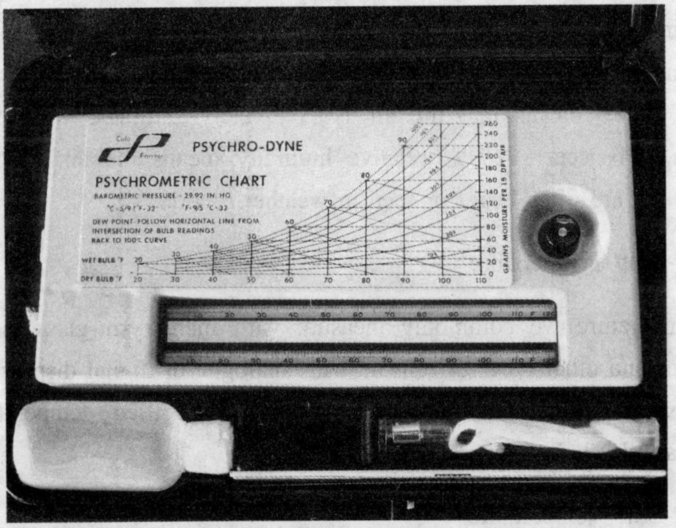

Figure 2-1. Dry-bulb and wet-bulb temperatures can be measured using a psychrometer and the relative humidity can be read from the psychrometric chart

2.1.1 Humidity Ratio

The air we breathe is a mixture of dry air and moisture, often called moist air. Humidity ratio, or moisture content, is the ratio of moisture mass (water vapor) to dry air mass in an air sample. A humidity ratio of 0.008, for example, indicates that a one kilogram mass of dry air contains eight grams of moisture (the ratio is expressed in kg/kg).

The air's ability to hold moisture depends on its temperature. The higher its temperature, the more moisture air can hold. When air can hold no more moisture at a given temperature, it is said to be saturated. When the saturation point is exceeded, condensation occurs. For example, when room air meets a cold bottle from a refrigerator, condensation occurs on the bottle surface. The bottle's low temperature reduces the temperature of air surrounding the bottle and this air falls below the saturation point. Similar situations occur when cold, outside air meets the warm, moist air inside pig buildings.

2.1.2 Relative Humidity

Relative humidity is the ratio of actual moisture content to the saturated moisture content of air. It is a measure of the extent to which air is saturated with moisture. Relative humidity is most often used to judge the "wetness" of the air. A relative humidity of 100% means that the air is completely saturated with moisture and condensation will occur. A 0% relative humidity means the air is dry. Relative humidity reports are often heard or seen in weather reports.

2.1.3 Dry-Bulb Temperature

The temperature we commonly measure with mercury-in-glass thermometers, thermocouples and other types of sensors with analogue or digital displays is properly referred to as dry-bulb temperature. Unless otherwise specified, temperatures used in this manual are dry-bulb temperatures.

2.1.4 Wet-Bulb Temperature

The temperature at which water evaporates and brings air to saturation without heat loss or gain (adiabatically) is called the wet-bulb temperature. For air to become saturated adiabatically at a given dry-bulb temperature and humidity ratio, the pressure must remain constant.

2.1.5 Dew-Point Temperature

Dew-point temperature is the temperature at which condensation occurs on a solid surface. For condensation to occur, moist air must be saturated with the same moisture content and at the same pressure as the sample air. For example, when you enter a barn in cold weather, your eyeglasses will fog up because they are below the room air's dew-point temperature.

2.1.6 Enthalpy

Enthalpy is the internal energy content of moist air, expressed in kJ/kg (1 kJ = 1/3 600 kW · h). The enthalpy of moist air is the sum of two energy components: sensible heat and latent heat. Sensible heat (SH) is the energy content of the dry air in any moist air sample. Latent heat (LH), in livestock ventilation application, is considered the energy content of the water vapor in that same air sample.

2.2 Psychrometric Chart

Commonly used psychrometric charts are plotted in Figure 2-2 (ASHRAE, 2005). In a psychrometric chart, the horizontal axis represents the dry-bulb temperature and the vertical axis represents the humidity ratio. Moist air properties contained in a psychrometric chart include relative humidity, wet-bulb temperature, dew-point temperature, enthalpy, and specific volume. The physical and thermal properties of moist air can be determined from any two known properties. Although a psychrometric chart will be different with a different altitude from the sea level, the chart in Figure 2-2 is considered sufficiently accurate for swine housing application in most areas. A different chart is recommended for areas with a high altitude (typically 1 000 meters above sea level).

The following example will help you determine the psychrometric properties of a moist air sample taken from a barn.

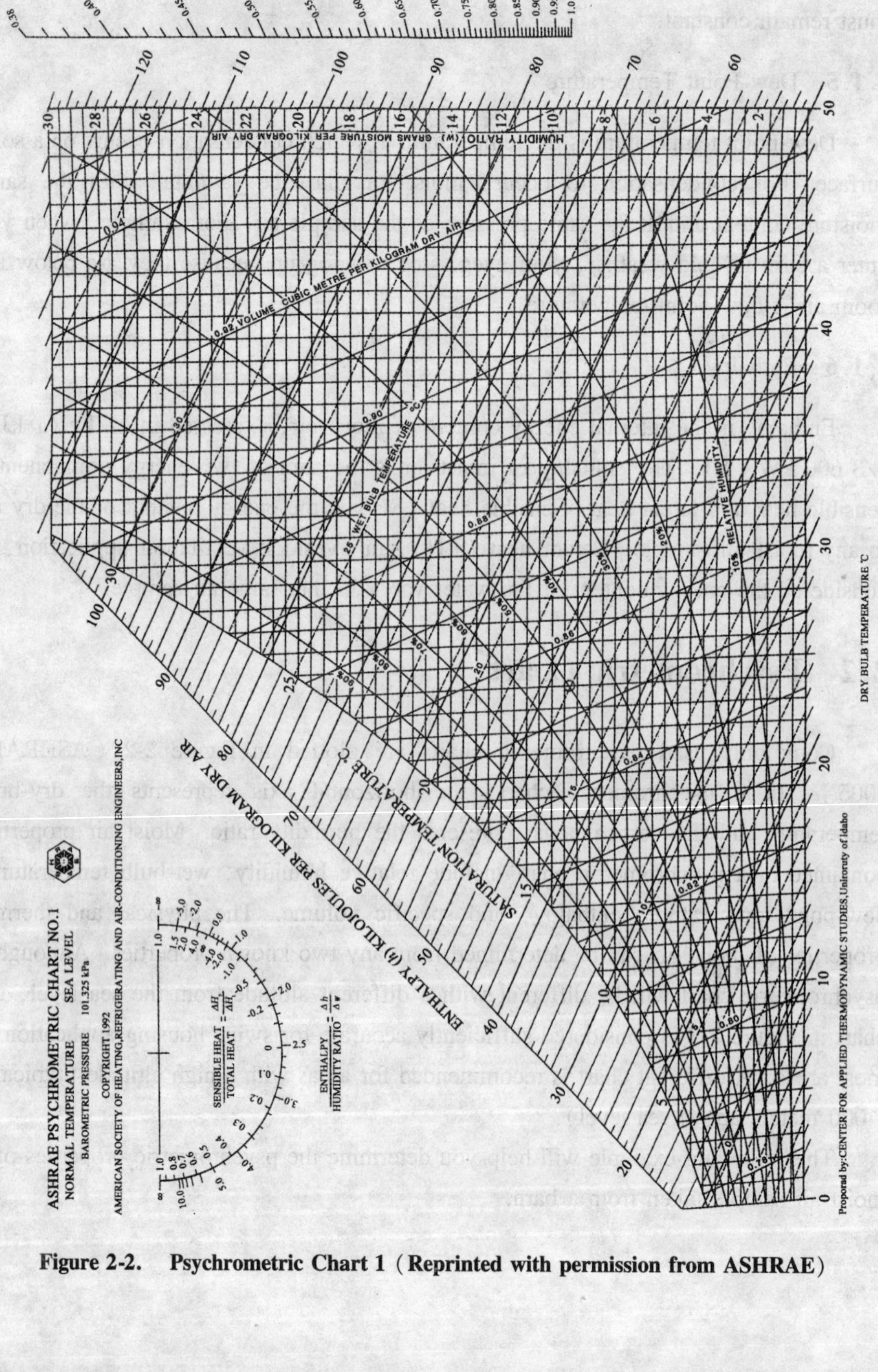

Figure 2-2. Psychrometric Chart 1 (Reprinted with permission from ASHRAE)

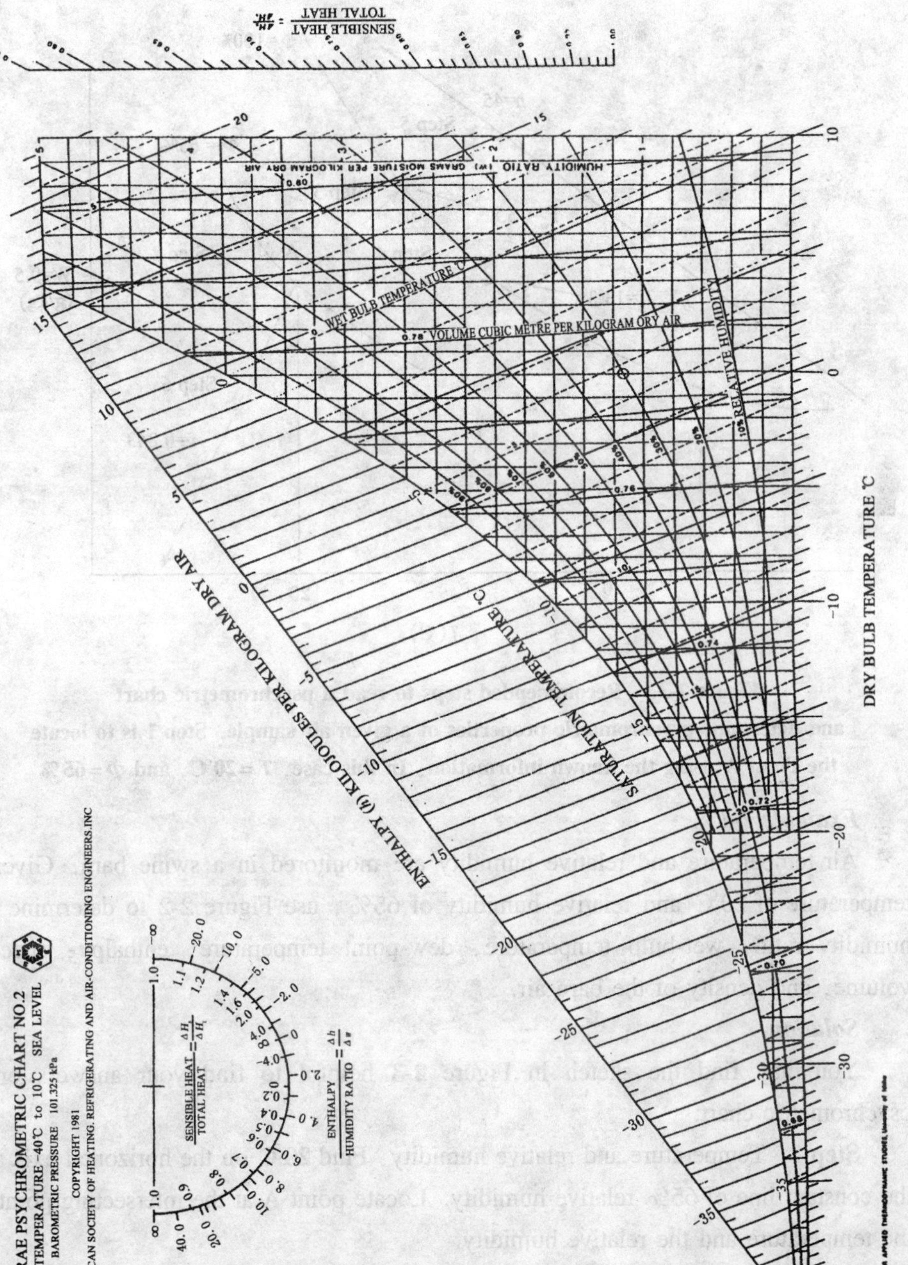

Figure 2-2. Continued. Psychrometric Chart 2 (Reprinted with permission from ASHRAE)

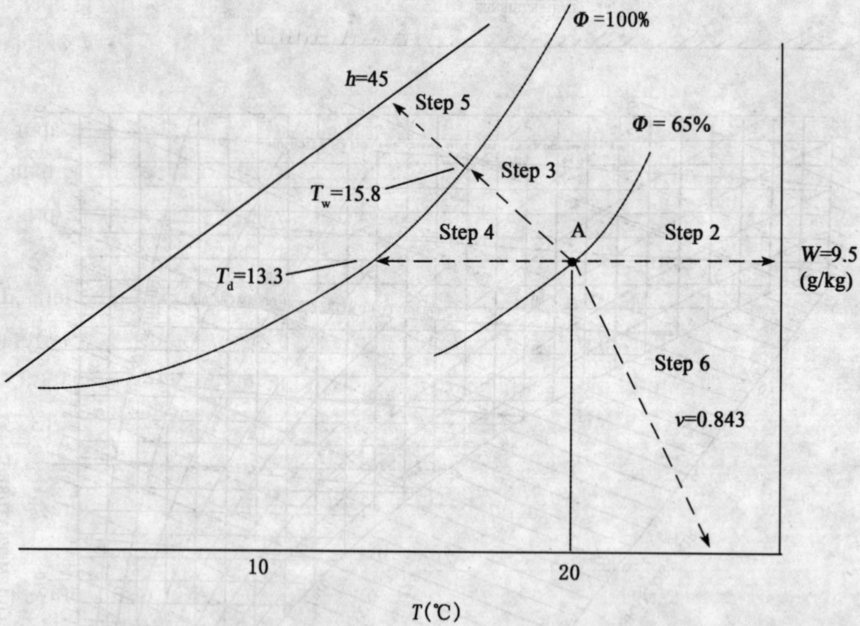

Figure 2-3. Recommended steps to read a psychrometric chart and determine psychrometric properties of a given air sample. Step 1 is to locate the Point A using the known information: In this case, $T = 20$℃, and $\Phi = 65\%$

Example 2-1

Air temperature and relative humidity are monitored in a swine barn. Given a temperature of 20℃ and relative humidity of 65%, use Figure 2-2 to determine the humidity ratio, wet-bulb temperature, dew-point temperature, enthalpy, specific volume, and density of the barn air.

Solution

You may find the sketch in Figure 2-3 helpful to find your answers on a psychrometric chart.

Step 1. Temperature and relative humidity: Find 20℃ on the horizontal axis and the constant line of 65% relative humidity. Locate point A at the intersecting point of the temperature and the relative humidity.

Step 2. Humidity ratio: Starting at point A and moving to the right, draw a line parallel to the horizontal axis (temperature) and intersect the vertical axis (humidity ratio). The intersection point at the vertical axis gives the humidity ratio of 0.009 5.

Step 3. Wet-bulb temperature: Starting at point A and moving to the upper left, draw a line parallel to the wet-bulb temperature line and intersect the 100% relative humidity line. The intersection point indicates the wet-bulb temperature of 15.8℃. This is the temperature at which this air sample could be cooled by evaporative cooling. This phenomenon of evaporative cooling is similar to the cooling that you experience on a hot day when water evaporates off your skin. It is also the principle involved in spray cooling of pigs.

Step 4. Dew-point temperature: Starting at point A and moving to the left, draw a line parallel to the horizontal axis (temperature) and intersect the 100% relative humidity line. The intersection point gives the dew-point temperature of 13.3℃. If the air in this example comes into contact with a surface which is lower than 13.3℃, condensation will occur.

Step 5. Enthalpy: Starting at point A and moving to the upper left, draw a line parallel to the enthalpy line and read the enthalpy ~45 kJ/kg.

Step 6. Density: Starting at point A and moving to the lower right, draw a line parallel to the specific volume line. The line is about one quarter between 0.84 and 0.85, therefore, the specific volume of the barn air is 0.8425 m^3/kg. Since density is the reciprocal of the specific volume, the density of the barn air is 1/0.842 5, which is 1.187 kg/m^3.

From this example, you can see that with any two known properties of a given air sample, you can locate a point on a psychrometric chart. Based on that point and repeating the steps in Example 2-1, you can determine other psychrometric properties.

An important fact that can be determined from a psychrometric chart is that although the relative humidity is high, the air's moisture content in cold weather is very low and has little effect on barn air. For example, a saturated (100% relative humidity) outside air at temperature −20℃ will only have a 10% relative humidity when it is heated to 10℃. For many practical housing considerations, moisture content of outside air is negligible when the temperature is below −15℃.

For building environment control, temperature ranges from −40℃ to 40℃. Two sets of psychrometric charts are provided in the back of this text: (a) normal temperature, 0℃ to 50℃ and (b) a cold temperature, −40℃ to 10℃. (*Reprinted with permission from the American Society of Heating, Refrigeration and Air-Conditioning Engineers [ASHRAE], 2005.*)

2.3 Thermal Comfort Zone

A thermal comfort zone is a range of effective temperatures that provide a comfortable environment for people. In some literature it may also be referred to as the thermal comfort control zone. The effective temperature is a function of dry-bulb temperature, relative humidity, irradiative temperature, and air speed. It is defined as the temperature at 50% relative humidity that yields the same total heat loss from the skin as for the actual environment. Similar to a wind chill factor, it is a more relevant measurement of "coldness." Additional scientific definitions and calculations on effective temperatures can be found in the *ASHRAE Handbook of Fundamentals* (*2005*).

The American Society of Heating, Refrigeration and Air-Conditioning Engineers (ASHRAE) define the comfort zone for summer and winter seasons, taking the energy conservation into consideration (Figure 2-4).

2.4 Case Study: Psychrometrics Classroom Exercise

Objectives: To measure the dry-bulb and wet-bulb temperatures of the classroom air and to determine other psychrometric properties of the room air using a psychrometric chart.

Step 1: Find your partner next to you (In case of an odd number, work by yourself or join the nearest pair).

Step 2: One volunteer in the class measures the dry-bulb and wet-bulb temperature. Put the measurements into the Answer Table below.

Step 3: Work in pairs to find other psychrometric properties of the classroom air using the chart and put answers in the column of "Yourself".

Step 4: Establish work teams:

- Students in adjacent rows will face each other to form a work team.
- Check your results within the team-the team should have unanimous answers.
- Put the team answer in the column of "Team".
- A team leader is elected to report later.

Step 5: Team report of the results. Each team gives an answer of one property designated by the instructor.

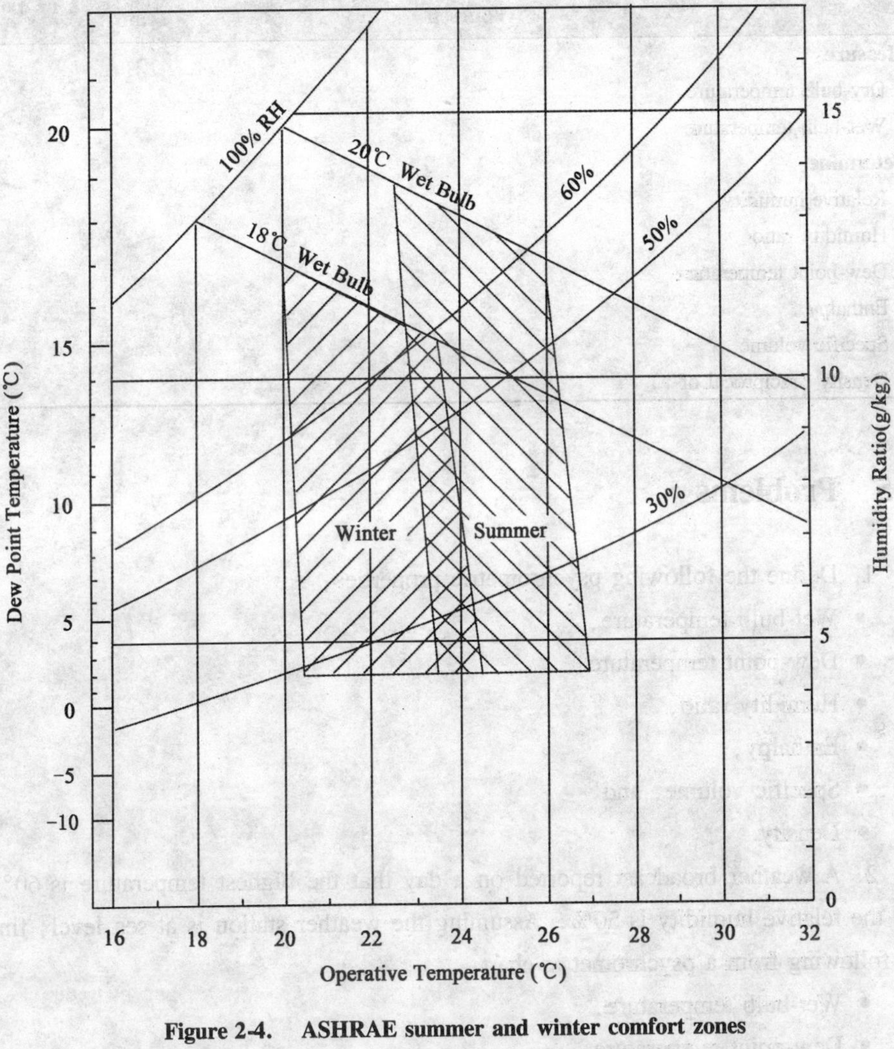

Figure 2-4. ASHRAE summer and winter comfort zones (acceptable ranges of operative temperature and humidity for people in typical summer and winter clothing during primarily sedentary activity)

Answer Table

	Results	
	Yourself	**Team**
Measure:		
Dry-bulb temperature		
Wet-bulb temperature		
Determine:		
Relative humidity		
Humidity ratio		
Dew-point temperature		
Enthalpy		
Specific volume		
Density (reciprocal of v)		

2.5 Problems

1. Define the following psychrometric properties:

- Wet-bulb temperature,
- Dew-point temperature,
- Humidity ratio,
- Enthalpy,
- Specific volume, and
- Density.

2. A weather broadcast reported on a day that the highest temperature is 60°F and the relative humidity is 50%. Assuming the weather station is at sea level, find the following from a psychrometric chart:

- Wet-bulb temperature,
- Dew-point temperature,
- Humidity ratio,
- Enthalpy,
- Specific volume, and
- Density.

3. Use the psychrometric chart to determine how much water vapor must be

added to 5 kg of air at $-15℃$ and 50% relative humidity, if it is to be conditioned to 25℃ and 30% relative humidity.

4. A room contains 100 kg of air which is initially at 15℃ and 60% of relative humidity. After a 1.5 kW baseboard electric heater was turned on in the room for 10 minutes, find:

a. How much heat was added into the room air?

b. What are the temperature and relative humidity of the air sample now?

5. Using sensible heating. A room contains 100 kg of moist air is initially at 10℃ and 70% relative humidity. Find:

a. How much heat is needed to heat the room air to 20℃?

b. What is the relative humidity now?

c. Sketch this sensible heating process.

Module 3: Psychrometric Processes

In tempering the air to achieve desirable conditions, temperature and relative humidity in particular, one or more of the psychrometric processes need to be applied. All the heating, ventilation and air-conditioning (HVAC) equipment or systems are designed to execute one or more of the psychrometric processes. There are five basic psychrometric processes in heating, ventilation and air-conditioning:

- Sensible heating or cooling
- Sensible heating with humidifying
- Sensible cooling with dehumidifying
- Evaporative cooling
- Adiabatic mixing

3.1 Sensible Heating or Cooling

Sensible heating or cooling refers to adding heat to or removing heat from the air without changing its humidity ratio. For example, adding heat to a room using an electrical heater is a process of sensible heating; and when the electrical heater is turned off, the electric heater is cooled to room temperature which is a sensible cooling process. The sensible heating and/or cooling process is represented in a psychrometric chart as a straight line parallel to the temperature axis. In Figure 3-1, the air psychrometric properties are changed between points 1 and 2. During the sensible heating process, air temperature changes from T_1 to T_2, and the sensible heat in the air increases from h_{s1} to h_{s2}, accordingly. During the sensible cooling, the temperature is reduced from T_2 to T_1, and the sensible heat is reduced from h_{s2} to h_{s1}, respectively.

The change of sensible heat to each kg of air, Δh_s, can be read from the chart or calculated from the difference of the temperature change.

$$\Delta h_s = h_{s2} - h_{s1} \tag{3-1}$$

Or

$$\Delta h_s = c\Delta T = 1.006(T_2 - T_1) \tag{3-2}$$

Because the sensible heat per kg of air is proportional to the temperature

$$h_s = cT \tag{3-3}$$

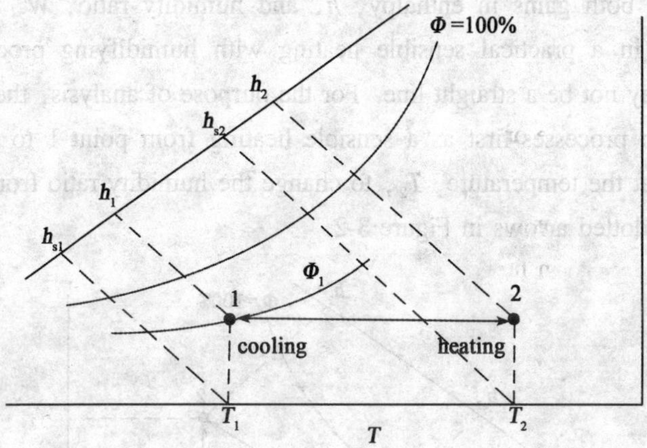

Figure 3-1. Schematic of a process of sensible heating or cooling

For the total sensible heat rate, q_s, added by an air flow rate, m, to cause the change from point 1 to point 2 (Figure 3-1), will be

$$q_s = m(h_{s2} - h_{s1}) \tag{3-4}$$

Where

c = 1.006 kJ/(kg · ℃), is the heat capacity of dry air

h_s = sensible enthalpy, kJ per kg of dry air

q_s = rate of sensible heat change, kJ/s

m = rate of air flow, kg/s

T = temperature, ℃

Note: When air changes from point 1 to point 2 in a sensible heating process, it changes temperature, enthalpy (sensible heat portion) and specific volume. It does not change the humidity ratio but changes the relative humidity.

3.2 Sensible Heating with Humidification

Sensible heating with humidifying process is to add both sensible heat and moisture to the air, usually simultaneously. This process will change all psychrometric properties of the air. For example, when a kettle of water is boiled on a stove, the room air changes. The boiled water and the kettle add both heat and steam to the room air, thus the process is heating and humidifying. The initial and final state of a sensible heating with humidifying process is described in a psychrometric chart as a straight line with both gains in enthalpy, h, and humidity ratio, W. It should be pointed out that in a practical sensible heating with humidifying process between points 1 and 2 may not be a straight line. For the purpose of analysis, the process can be treated as two processes-first as a sensible heating from point 1 to point 2 then adding moisture at the temperature, T_2, to change the humidity ratio from W_1 to W_2, as shown by the dotted arrows in Figure 3-2.

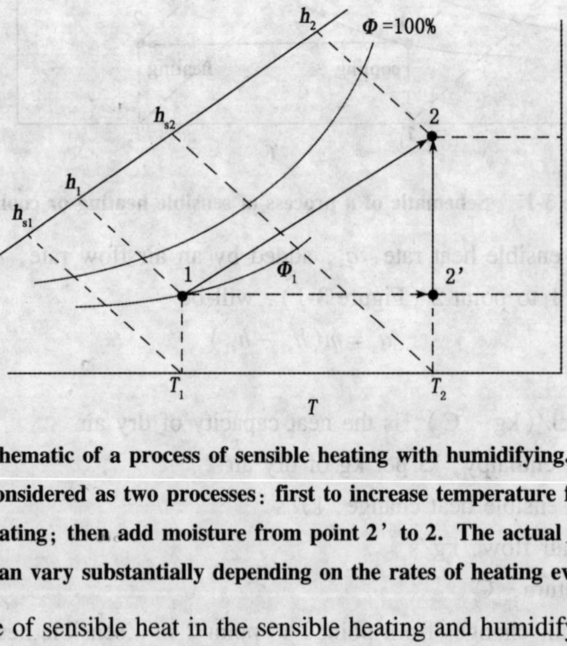

Figure 3-2. Schematic of a process of sensible heating with humidifying. Analytically, the process can be considered as two processes: first to increase temperature from points 1 to 2' as a sensible heating; then add moisture from point 2' to 2. The actual route from points 1 to 2 can vary substantially depending on the rates of heating evaporation

The change of sensible heat in the sensible heating and humidifying process is the same as that of the sensible heating process (Equation 3-2). The change of moisture content, or humidity ratio, ΔW, can be read from the chart, or calculated from the difference of the humidity ratios at states 1 and 2 (points 1 and 2 in Figure 3-2).

$$\Delta W = W_2 - W_1 \tag{3-5}$$

The latent heat portion of the enthalpy, h_w, can be determined by the difference between the enthalpy at points 2 and 2' on the psychrometric chart

$$\Delta h_w = h_2 - h_{2'} \tag{3-6}$$

or calculated using Equation 3-7.

$$\Delta h_w = \Delta W(2\ 501 + 1.805T) \tag{3-7}$$

Where

h_w = latent heat portion of enthalpy, kJ/kg of dry air

h_2, $h_{2'}$ = enthalpy at points (states) 2 and 2', kJ/kg of dry air

T = air temperature at the final state, ℃

For the total latent heat rate, q_w, added by an air flow rate, m, to cause the change from point 1 to point 2 (Figure 3-2), will be

$$q_w = m\Delta h_w \tag{3-8}$$

Where

h_w = latent enthalpy, kJ per kg of dry air

q_w = rate of sensible heat change, kJ/s

m = rate of air flow, kg/s

From Equation 3-2 and Equation 3-7, the total enthalpy change of the air from point 1 to point 2 is

$$\Delta h = \Delta h_s + \Delta h_w = 1.006T + \Delta W(2\ 501 + 1.805T) \tag{3-9}$$

The enthalpy of the air at temperature, T, and humidity ratio, W, is

$$h = h_s + h_w = 1.006T + W(2\ 501 + 1.805T) \tag{3-10}$$

3.3 Sensible Cooling with Dehumidification

The sensible cooling with dehumidifying process is to reduce the temperature and remove moisture in the air. Unlike the sensible heating and cooling, the humidifying process can not be reversed to achieve dehumidifying. Instead, as shown in Figure 3-3, the sensible cooling with dehumidifying process goes sensible cooling first from the initial state (point 2) until reaching the saturation line at point 1, at which point the temperature reaches the dew-point temperature and Φ = 100%. The water vapor at or below the dew-point temperature in the air is condensed into liquid and separated from the air along the saturation line until reaching the designed cooling temperature. Sensible cooling with dehumidifying is the most popular process in air-conditioning, including those used in homes and cars. Have you noticed a water spot under the engine of your vehicle after stopping by a gas station during a hot summer day? This water spot is a result of the sensible cooling with dehumidifying of the air-conditioner.

Note: In sensible cooling with dehumidifying (Figure 3-3), the initial state is point 2 and the final state is point 1. The changes of temperature, humidity ratio and enthalpy during the sensible cooling with dehumidifying can be determined using Equation 3-1 to Equation 3-10.

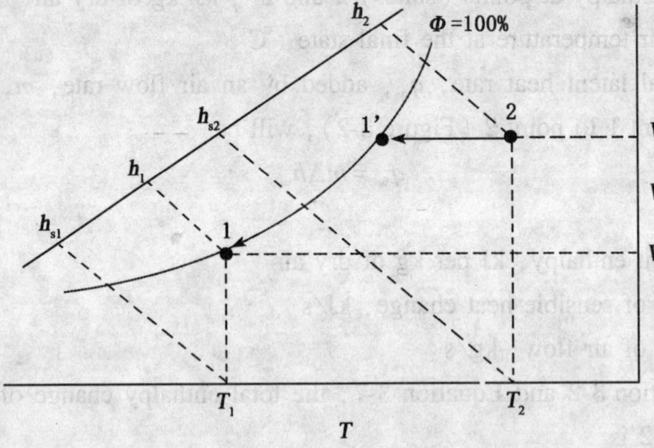

Figure 3-3. Schematic of a process of sensible cooling with dehumidifying

3.4 Evaporative Cooling

The evaporative cooling process is to reduce the dry-bulb temperature by evaporating water into water vapor adiabatically. Because it is adiabatic, there is no net heat gain or loss during the process. Rather, it converts the sensible heat into latent heat. Because dry-bulb temperature is directly proportional to sensible heat, when the sensible heat is reduced, the dry-bulb temperature is reduced, hence the air is cooled. Sweating is an evaporative cooling process. Many animal buildings in hot climates use evaporative cooling systems instead of sensible cooling with dehumidifying. As shown in Figure 3-4, the evaporative cooling process starts from initial point 1 to point 2 along the enthalpy line, thus there is no change in energy (adiabatic).

It is obvious that evaporative cooling has a limit-the best it can do is to reach the wet-bulb temperature of the initial state (point 1). If the final state (point 2) reaches the saturation line, the evaporative cooling process reaches its maximum potential; often we say the process has a relative 100% efficiency. The relative efficiency is "relative" to the initial point 1. However, if the final state (point 2) is somewhere between point 1 and the saturation line, the process relative efficiency is less than 100%. The process relative efficiency, η, can be determined by

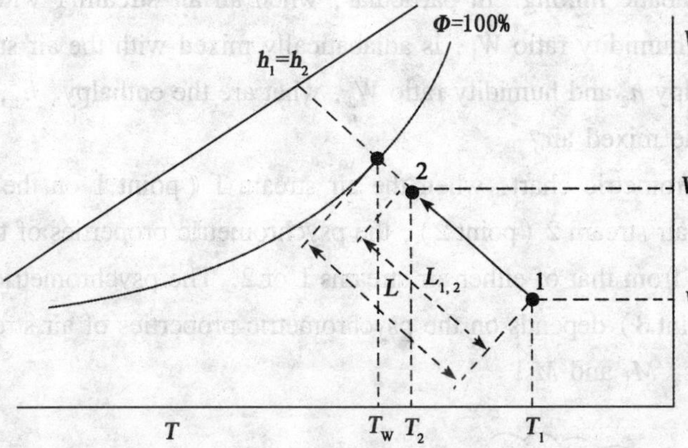

Figure 3-4. Schematic of a process of evaporative cooling

$$\eta = \frac{L_{1\text{-}2}}{L} \times 100\% \tag{3-11}$$

Where

$L_{1\text{-}2}$ = distance between points 1 and 2

L = distance between point 1 and the saturation line along the enthalpy line

From Figure 3-4, it is clear that an evaporative cooling process can only be effective for a hot and dry climate, i. e., have a large L or ΔT potential. If the air is hot and also humid, point 1 will be close to the saturation line, thus leaving little room for evaporative cooling. Even if the process has a high relative efficiency, the ΔT that can be achieved will be small.

3.5 Adiabatic Mixing of Two Moist Air Streams

Adiabatic mixing is a process where two streams of moist air are mixed adiabatically (Figure 3-5). There is no external energy gain or loss during the mixing process. Adiabatic mixing exists everywhere in our daily life. When the hot air comes from a heater into a room airspace, the hot air is adiabatically mixed with the cooler room air; when you breathe, the moist warm air (gas) is adiabatically mixed with the ambient air; and when a window air-conditioner is operating, the cooling air from the air-conditioner is adiabatically mixed with the room air.

The question of interest is what the new psychrometric properties of the mixed air are after the adiabatic mixing. In particular, when an air stream 1 with a mass M_1, enthalpy h_1 and humidity ratio W_1, is adiabatically mixed with the air stream 2 with a mass M_2, enthalpy h_2 and humidity ratio W_2, what are the enthalpy, h_3, and humidity ratio, W_3, of the mixed air?

On a psychrometric chart, when the air stream 1 (point 1 on the chart) mixes with a different air stream 2 (point 2), the psychrometric properties of the mixture air will be different from that of either air streams 1 or 2. The psychrometric properties of the mixture (point 3) depends on the psychrometric properties of air streams 1 and 2, and their masses, M_1 and M_2.

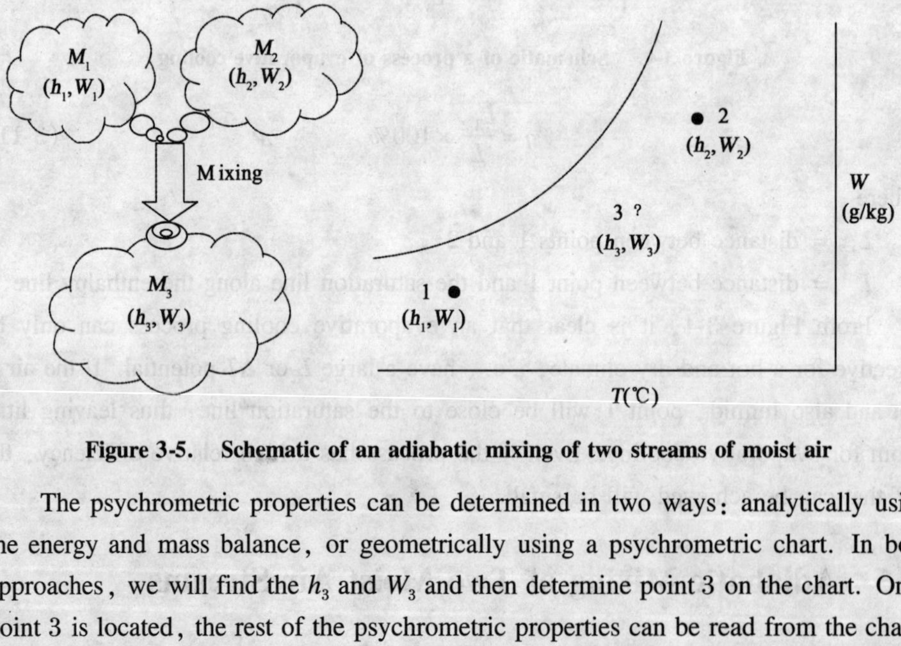

Figure 3-5. Schematic of an adiabatic mixing of two streams of moist air

The psychrometric properties can be determined in two ways: analytically using the energy and mass balance, or geometrically using a psychrometric chart. In both approaches, we will find the h_3 and W_3 and then determine point 3 on the chart. Once point 3 is located, the rest of the psychrometric properties can be read from the chart. It is worth mentioning that any two psychrometric properties of the mixed air could be used to determine point 3. However, because the enthalpy and humidity ratio (moisture content) directly represents the energy and water vapor mass in the mixed air, it is relatively more direct to apply mass and energy conservation equations.

Analytical Method: When two air streams mix together, mass and energy conserve.

From energy conservation:

$$M_1 h_1 + M_2 h_2 = M_3 h_3 \tag{3-12}$$

From mass conservation, neglecting the change between latent and sensible heat:

For water vapor: $M_1 W_1 + M_2 W_2 = M_3 W_3$ (3-13)

For total mass: $M_1 + M_2 = M_3$ (3-14)

Substituting Equation 3-14 into Equation 3-12 and Equation 3-13, gives:

$$h_3 = \frac{M_1 h_1 + M_2 h_2}{M_1 + M_2} \tag{3-15}$$

$$W_3 = \frac{M_1 W_1 + M_2 W_2}{M_1 + M_2} \tag{3-16}$$

Although the above discussions are based on mass, it is equally applicable to be based on flow rates at steady-state. In practice, mass flow rate, m, in kg of air per second or volumetric flow rate, Q, in cubic meter of air per second, are commonly used. In either case, mass and energy conservation still holds. Mass M, can be substituted by mass flow rate, m, or volumetric flow rate, Q/v (because $m = Q/v$), in Equation 3-12 to Equation 3-20.

Geometrical Method on a Chart: (find point 3 and then determine h_3 and W_3 [Figure 3-6]).

Rewrite Equation 3-15 and Equation 3-16,

$$h_3 = \frac{M_1}{M_1 + M_2} h_1 + \frac{M_2}{M_1 + M_2} h_2 \tag{3-17}$$

$$W_3 = \frac{M_1}{M_1 + M_2} W_1 + \frac{M_2}{M_1 + M_2} W_2 \tag{3-18}$$

Here h_3 is linear to h_1 and h_2 and W_3 is linear to W_1 and W_2. Therefore, point 3 must be located on the straight line connecting point 1 and point 2 on a psychrometric chart. The ratio of M_1 to M_3 or M_2 to M_3 determines where point 3 is.

Let's denote the distances between points 1 and 2, 2 and 3, and 1 and 3 as L_{1-2}, L_{2-3} and L, respectively, and then Equation 3-17 and Equation 3-18 can be rewritten as:

$$L_{1-3} = \frac{M_2}{M_1 + M_2} L \tag{3-19}$$

$$L_{2-3} = \frac{M_1}{M_1 + M_2} L \tag{3-20}$$

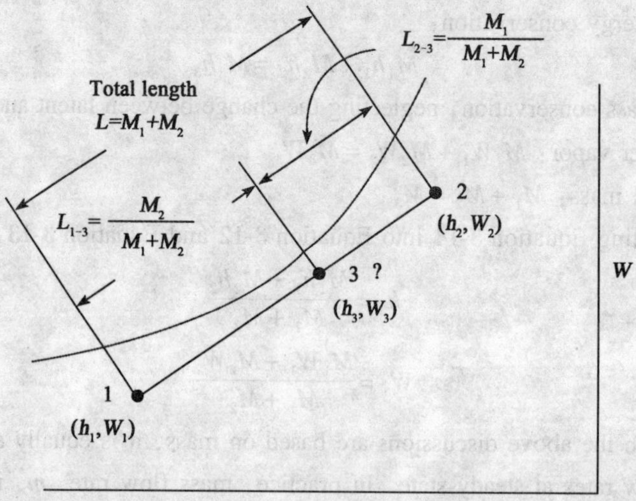

Figure 3-6. Geometric determination of mixed air psychrometric properties (point 3). Point 3 must be located on the straight line connecting Points 1 and 2, and the distances are proportional to the masses of the air streams 1 and 2

Example 3-1

When 10 kg of air from an air-conditioner at 15℃ and 100% relative humidity mixed with 30 kg of room air at 28℃ and 50% relative humidity, calculate the enthalpy and humidity ratio of the mixed air, verify the answer using the geometric method by sketching on a psychrometric chart. The process is under standard atmospheric conditions.

Solution

From the given properties and a psychrometric chart, we have

M_1 = 10 kg, Based on T_1 = 15℃ and Φ_1 = 100%

W_1 = 0.010 7 kg/kg

h_1 = 42 kJ/kg

M_2 = 30 kg, Based on T_2 = 28℃ and Φ_2 = 50%

W_2 = 0.011 8 kg/kg

h_2 = 58 kJ/kg

$$h_3 = \frac{M_1}{M_1+M_2}h_1 + \frac{M_2}{M_1+M_2}h_2 = \frac{10}{10+30}\times 42 + \frac{30}{10+30}\times 58 = 54\left(\frac{\text{kJ}}{\text{kg}}\right)$$

$$W_3 = \frac{M_1}{M_1 + M_2} W_1 + \frac{M_2}{M_1 + M_2} W_2$$

$$= \frac{10}{10+30} \times 0.010\ 7 + \frac{30}{10+30} \times 0.011\ 8 = 0.011 \left(\frac{\text{kg of water}}{\text{kg of air}}\right)$$

Using a psychrometric chart, we can locate point 3 and read the new temperature and relative humidity of the mixed air:

$$T_3 = 24.8℃$$

$$\Phi_3 = 57\%$$

3.6 Problems

1. On a summer day, temperature reaches 35℃ and relative humidity 50%. Find:

a. What is the possible lowest temperature that can be achieved when evaporative cooling is used?

b. What is the actual temperature that can be achieved if the evaporative cooling system has an efficiency of 80%?

c. Sketch this cooling process.

2. The outside air is at 35℃, 60% relative humidity (RH) is fed into an air conditioner. After cooling with the dehumidification process, the air exits the air conditioner at 15℃ and 100% RH. Using the psychrometric chart, find:

a. What is the humidity ratio after the air is cooled?

b. How much moisture is removed per kg of dry air?

c. Sketch the process.

3. Determine the temperature and relative humidity of a mixed air resulting from an adiabatic process where 20 kg of air at 20℃ and 80% relative humidity is mixed with 4 kg of air at 30℃ and 50% relative humidity. Sketch the process.

4. In a refrigeration unit, 2.8 m^3/s moist air at 14℃ temperature, 85% relative humidity enters the unit. The air is cooled by the refrigeration process to 4℃ and there is no reheating. Using the psychrometric chart, sketch the process, and determine

a. The minimum electrical capacity of the refrigeration unit, in kW.

b. The rate of water removal from the air.

c. Volumetric flow rate (at 4℃ and 100% RH) of the air leaving the refrigerator.

5. A product evaluation chamber is located within a large air-conditioned building. The temperature and relative humidity within the chamber are maintained constant at 22℃ and 50% relative humidity. However, psychrometric properties of the chamber air vary with the atmosphere pressure change. From the local weather record, atmosphere pressure can be as low as 90 kPa or as high as 105 kPa; find in the chamber the ranges (i. e., minimum and maximum) of:

a. Water content.

b. Dew-point temperature.

c. Enthalpy.

d. Sensible heat content.

e. Latent heat content.

f. Specific volume.

g. Density.

3.7 Laboratory Exercise: Psychrometrics

Instrumentation: Psychrometer

Procedures:

1. Hands-on experience to measure the dry-bulb temperature and wet-bulb temperature of the room air.

2. Determine other psychrometric properties:

- Relative humidity,
- dew-point temperature,
- humidity ratio,
- enthalpy,
- sensible heat (kJ), and
- latent heat (kJ).

In this step, assume sea level atmospheric pressure and use a standard psychrometric chart.

3. Assume the room contains 100 kg of moist air (as defined in the questions above). The room has 15 people in it and the room is sealed and well insulated, i.

e., there is no ventilation and no heat loss. Each person gives out 100 W of sensible heat and 40 g/h of moisture. After 10 minutes, find:

- Total sensible heat (kJ) added into the room.
- Total latent heat (kJ) added into the room.
- Temperature in the room.
- Relative humidity in the room.

4. Find out if there is any problem in the room. If there is, offer your solutions to solve the problem. (Depending on the nature of the problem in the room, solutions may include heating, cooling, adiabatic mixing, and air cleaning).

Conversions you may need:

Temperature: $℃ = \dfrac{℉ - 32}{1.8}$

Heat and power: 1 kW = 1 kJ/s = 3 600 kJ/h

Module 4: Calculation of Psychrometric Properties

All discussions in previous sections about the psychrometric properties and psychrometric chart are based on one given atmospheric pressure, typically at sea level, i. e., P = 101. 325 kPa. However, psychrometric properties can be substantially different at high sea level. For example, in Tibet, water starts to boil at about 70℃. When the atmospheric pressure changes, other psychrometric properties change and the psychrometric chart becomes a 3-dimensional chart.

Pressure varies smoothly from the earth's surface to the top of the mesosphere. Although pressure changes with the weather, NASA① has averaged the conditions for all parts of the earth year-round. The following is a list of air pressures (as a fraction of one atmospheric pressure, 1 bar or 101. 325 kPa) with the corresponding average altitudes. Table 4-1 gives a rough idea of air pressure at various altitudes.

There are two different equations for computing the average pressure at various height regimes below 86 km (or 278 400 ft.) Equation 4-1 is used when the value of standard temperature lapse rate is not equal to zero and Equation 4-2 is used when the standard temperature lapse rate equals zero. The temperature lapse rate is defined as the negative of the rate of change in an atmospheric temperature with altitude (http://en. wikipedia. org/wiki/Atmospheric_pressure). For building environmental control purposes, Equation 4-1 is sufficiently accurate.

$$P = P_b \times \left[\frac{T_b}{T_b + L_b \times (h - h_b)}\right]^{\frac{g_0 \times M}{R^* \times L_b}} \tag{4-1}$$

$$P = P_b \times \exp\left[\frac{-g_0 \times M \times (h - h_b)}{R^* \times T_b}\right] \tag{4-2}$$

Where

P_b = Static pressure (Pascals)

T_b = Standard temperature, K

①National Aeronautics and Space Administration.

L_b = Standard temperature lapse rate (Kelvins per m)

h = Height above sea level (meters)

h_b = Height at bottom of layer b (meters, e.g., h_1 = 11 000 m)

R^* = Universal gas constant: $8.314\ 32 \times 10^3$ N · m / (kmol · K)

g_0 = Standard gravity (9.806 65 m/s^2)

M = Molar mass of Earth's air (28.964 4 g/mol)

Table 4-1. Average atmospheric pressure at different altitudes

Altitude		Atmosphere Pressure
(ft)	(m)	(kPa)
0	0	101.325
500	153	99.49
1 000	305	97.63
1 500	458	95.91
2 000	610	94.19
2 500	763	92.46
3 000	915	90.81
3 500	1 068	89.15
4 000	1 220	87.49
4 500	1 373	85.91
5 000	1 526	84.33
6 000	1 831	81.22
7 000	2 136	78.19
8 000	2 441	75.22
9 000	2 746	72.40
10 000	3 050	69.64
15 000	4 577	57.16
20 000	6 102	46.61
25 000	7 628	37.65
30 000	9 153	30.13
35 000	10 679	23.93

Therefore, atmospheric pressure must be a known property to uniquely define the point on a psychrometric chart, together with two other properties (temperature and relative humidity for example). It is useful in some cases that one is required to calculate the local psychrometric properties in order to design systems related to air.

Mathematical analysis of psychrometric properties are in Appendix 2 (Albright, 1999).

The following case study will guide students in a classroom environment, either individually or as teams, to work out the calculation procedures of psychrometric properties at a specific atmospheric pressure.

4.1 Case Study: Calculation of Psychrometric Properties

4.1.1 Situation and Task Description

You are head of the design division which consists of five engineers in a large firm. Your division is equipped with basic instrumentation such as a psychrometer (measures temperature, relative humidity and wet-bulb temperature) and a barometer (measures atmosphere pressure). One day, the product evaluation division needs a chamber within which air properties should be precisely controlled. To control the chamber air, the psychrometric properties of the outside air must be determined, and then serve as input data for a control panel. Doug Pershing, Director of Research and Development, calls you and wants your division to provide him with the outside air properties and the needed procedures to obtain these properties … in as short a time as possible. He is in an extreme hurry to get the product evaluated to try and be ahead of the competitors.

- Temperature
- Wet-bulb temperature
- Relative humidity
- Water content
- Dew-point temperature
- Enthalpy
- Sensible heat
- Latent heat
- Specific volume
- Density

4.1.2 Approach

Task 1: Establish teamwork. You call a meeting immediately and describe the situation to the members in your division. Determine objectives and list tools (such as a calculator, etc.) that you need to achieve your objectives.

Task 2: Collect information. You have now clarified your objectives and have the necessary tools on hands. "*We need to have at least two psychrometric properties* *to start with.*" Rich suggested. Everybody agreed. Tom, a diligent and enthusiastic team player, volunteered to operate the psychrometer and measured the temperature as 22℃, wet-bulb temperature 14℃ and relative humidity 52%.

"*This is an easy job*," exclaimed Eric, "*I can find the rest of the properties now with a psychrometric chart.*" He pulled out a chart from his folder and started to work on it.

"*Wait a minute fellas.*" Susan, who has been quiet so far, interrupted, "*We can't use this chart. This chart is for standard atmospheric pressure, which is 101.325 kPa, at sea level only. We need to* ***calculate***, ***not read from the chart***, *those unknown psychrometric properties. I think we need to collect more information.*"

Finalize the tool list that you need to achieve your objectives.

Task 3: Determine the strategy. Your team agreed that to be able to calculate the rest of psychrometric properties, you need to know the atmospheric pressure. Tom, again, grabs a barometer and measures the atmospheric pressure (P), which is 95 kPa. Your team finally collected dry-bulb temperature, relative humidity, wet-bulb temperature, and atmospheric pressure. You also have all the necessary tools: a calculator and a book (Albright's book).

Dave, a clear-minded, task-oriented young engineer, suggests: "*We need to list all the variables, first the knowns and then unknowns, and then find the paths to find these unknowns.*" Your team members agreed with Dave and established the following table and calculated the unknowns:

Scenario 1: Known temperature, relative humidity and atmospheric pressure

(**All equation numbers in the following table refer to those in Appendix 2**)

Psychrometric Properties	Approaches (e. g., measured, equations used, etc.)	Medium Variables Needed
Temperature, T = 22℃	Measured using a psychrometer	
Relative humidity, Φ = 52%	Measured using a psychrometer	
Wet-bulb temperature, T_w = 14℃	Measured using a psychrometer, reiterate calculation, Equation 2-22 and Equation 2-23	
Atmosphere pressure, P_o = 95 kPa	Measured using a barometer	
Humidity ratio, W =		
Dew-point temperature, T_d =		

Scenario 1 Continued

Psychrometric Properties	Approaches (e. g. , measured, equations used, etc.)	Medium Variables Needed
Enthalpy, h =		
Sensible heat content, h_s =		
Latent heat content, h_w =		
Specific volume, v =		
Density, ρ =		

Your team has successfully completed the table and found all the necessary psychrometric properties needed. You can claim that "***We now can program this process in a computer and plot a local psychometric chart***!"

Scenario 2: Known temperature, wet-bulb temperature and atmospheric pressure (All equation numbers refer to the numbers in Appendix 2)

Psychrometric Properties	Approaches (e. g. , measured, equations used, etc.)	Medium Variables Needed
Temperature, T = 22℃	Measured using a psychrometer	
Wet-bulb temperature, T_w = 14℃	Measured using a psychrometer, reiterate calculation, Equation 2-22 and Equation 2-23	
Atmosphere pressure, P_o = 95 kPa	Measured using a barometer	
Humidity ratio, W	Using iteration, Equation 2-22 and Equation 2-23 to find W (computer program recommended)	
P_w	Equation 2-9	
P_{ws}	Equation 2-1	
Φ	Equation 2-2	
Dew-point temperature, T_d =	The same as the Scenario 1	
Enthalpy, h =		
Sensible heat content, h_s =		
Latent heat content, h_w =		
Specific volume, v =		
Density, ρ =		

Scenario 3: Known temperature, dew-point temperature and atmospheric pressure (All equation numbers refer to the number in Appendix 2)

Psychrometric Properties	Approaches (e. g. , measured, equations used, etc.)	Medium Variables Needed
Temperature, T = 22℃	Measured using a psychrometer	
Dew-point temperature, T_d = 10℃	Measured using a dew-point thermometer	
Atmosphere pressure, P_o = 95 kPa	Measured using a barometer	
Humidity ratio, W		
P_w		
P_{ws}		
Φ		
Dew-point temperature, T_d =		
Enthalpy, h =		
Sensible heat content, h_s =		
Latent heat content, h_w =		
Specific volume, v =		
Density, ρ =		

Key answers to Scenario 1 (All equation numbers refer to the numbers in Appendix 2)

Psychrometric Properties	Approaches (e. g. , measured, equations used, etc)	Medium Variables Needed
Temperature, T = 22℃	Measured using a psychrometer	
Relative humidity, Φ = 52%	Measured using a psychrometer	
Wet-bulb temperature, T_w = 14℃	Measured using a psychrometer (Calculation is iterative, Equation 2-22)	
Atmosphere pressure, P =95 kPa	Measured using a barometer	
Humidity ratio, W =	Equation 2-1 for saturated water vapor partial pressure (**Note**: That absolute temperature, K, should be used)	P_{ws}
	Equation 2-2 for actual water vapor partial pressure	
	Equation 2-9 for water content, or humidity ratio, W	P_w

Continued

Psychrometric Properties	Approaches (e. g., measured, equations used, etc)	Medium Variables Needed
Dew-point temperature, T_d =	Equation 2-17, for −60℃ to 0℃	
	Equation 2-18, for 0℃ to 70℃	
Enthalpy, h =	Equation 2-21	
Sensible heat, h_s =	Equation 2-21, first part of right side of equation	
Latent heat, h_w =	Equation 2-21, second part of right side of equation	
Specific volume, v =	Find gas constant for dry air, R_a = 287.055 J/(kg · K)	R_a
	Equation 2-16 for v. (**Note**: T is absolute temperature)	
Density, ρ =	Is the reciprocal of v	

Module 5: Basics of Heat Transfer

Heat transfer is the transition of thermal energy from a high temperature object (solid, liquid or gas) to a lower temperature object or its surroundings. Where there is a temperature difference between objects in proximity, heat transfer between them can never be stopped; it can only be slowed down.

In the simplest of terms, heat transfer is concerned with only two things: *temperature* and *flow of heat*. Temperature represents the potential of the thermal flow, just like water in a river, whereas heat flow represents the rate of thermal energy from place-to-place. On a microscopic scale, thermal energy is related to the kinetic energy of molecules. The greater a material's temperature, the greater the internal thermal energy, manifested both in linear motion (such as in fluid) or vibration modes (such as in solids).

Aside from temperature differences, several material properties affect the heat transfer. Such properties include thermal conductivities, specific heats, material densities, fluid velocities, fluid viscosities, and surface emissivities. In addition to the material properties, geometry of the object (from one-dimensional to three-dimensional) can affect heat transfer substantially. Moreover, the heat transfer could be considered at steady-state or transient. Heat transfer takes three modes: *conduction*, *convection and radiation*, individually or a combination in a particular heat transfer process. Considering all these factors, a solution to a heat transfer problem can be an involved process.

In this module, we will only discuss heat transfer under conditions of one-dimensional and steady-state. Even with these simplified conditions, we are still able to estimate the heat transfer rate (heating load or cooling load) through a building shelter under particular climate conditions.

5.1 Conduction Heat Transfer

Heat conduction is the thermal energy transfer from a high temperature object or region to a low temperature object or region through a solid, opaque medium.

From a microscopic point of view, conduction is a process that regions with greater molecular kinetic energy will pass their thermal energy to regions with less molecular energy through direct molecular collisions. In metals, a significant portion of the transported thermal energy is also carried by conduction-band electrons.

When a solid opaque object is subjected to two temperatures (Figure 5-1), heat conduction occurs from the high temperature region, T_h, to the low temperature region, T_l.

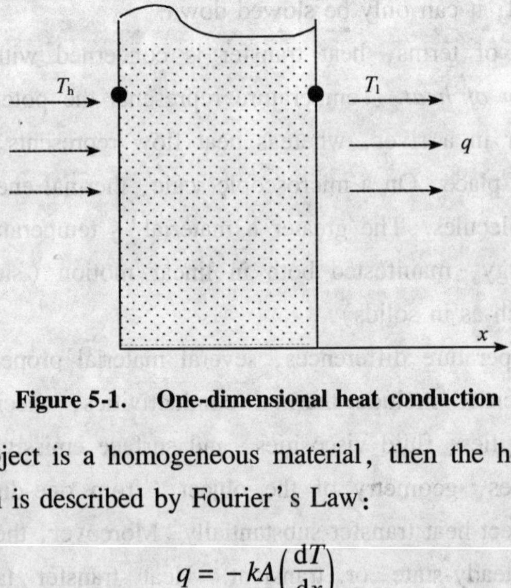

Figure 5-1. One-dimensional heat conduction

If the solids object is a homogeneous material, then the heat transfer rate, q, along the x-direction is described by Fourier's Law:

$$q = -kA\left(\frac{dT}{dx}\right) \tag{5-1}$$

Where

q = heat transfer rate, W

k = conductivity of the material, $\left(\frac{W}{m \cdot ℃}\right)$

A = heat transfer area, m^2

T = temperature, ℃

x = distance, m

The sign of "−" represents the heat transfer is always "down hill", from a high temperature region to a lower temperature region. The conductivity, k, is a thermal property of a given material, which can either be found in literature or can be measured if not readily available.

For steady-state process and homogenous material and when there is no heat sink or source, the temperature gradient is linear (Figure 5-2).

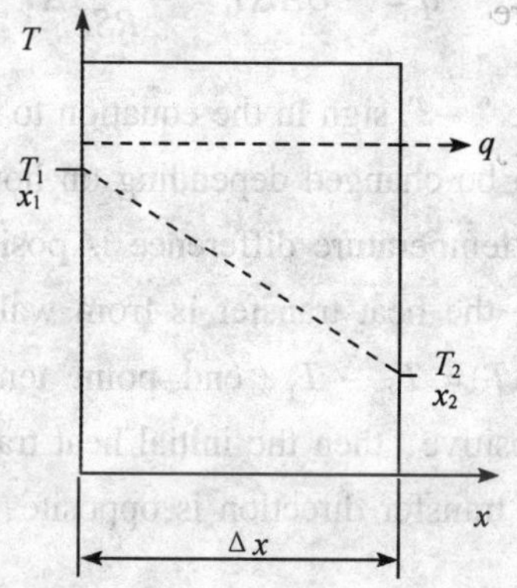

Figure 5-2. One-dimensional heat conduction through a homogeneous wall

Using a simple finite approximation for the temperature derivative, the Fourier Equation (Equation 5-1) becomes,

$$q = -kA\frac{\Delta T}{\Delta x} \tag{5-2}$$

Where

$$\Delta T = T_2 - T_1$$
$$\Delta x = x_2 - x_1$$

In the area of heat transfer for buildings, the material heat conductivity is usually converted to heat conductance, U, in $\left(\frac{\mathrm{W}}{\mathrm{m}^2 \cdot ℃}\right)$

$$U = \frac{k}{\Delta x} \tag{5-3}$$

or to thermal resistance, commonly called insulation value or *R*-value, *R*. The *R*-value for many insulation materials is usually represented in either English units ($\mathrm{h \cdot ft^2 \cdot ℉/Btu}$) and labeled *R* or in SI units ($\mathrm{℃ \cdot m^2/W}$) and labeled *RSI*. In this text, we will use *RSI* to denote the *R*-value of a material in SI units.

$$RSI = \frac{\Delta x}{k} = \frac{1}{U} \tag{5-4}$$

Then the conductive heat transfer in one-dimensional, through homogeneous material becomes

$$q = -UA\Delta T = -\frac{A}{RSI}\Delta T \tag{5-5}$$

Note: We still keep the " – " sign in the equation to be consistent with Fourier's Law. The negative sign can be changed depending on how the heat transfer direction is assumed, or whether the temperature difference is positive or negative. As shown in Figure 5-2, if we assume the heat transfer is from wall side 1 to side 2, then the temperature difference is $\Delta T = T_2 - T_1$ (end point temperature minus start point temperature). If the q is positive, then the initial heat transfer direction is correct; if q is negative, then the heat transfer direction is opposite, i. e., heat flows from wall side 2 to side 1.

In practice, U or R can be found in publications such as heat transfer textbooks and the *Fundamentals Handbook* (*ASHRAE*, *2005*). Thermal properties of typical building materials are listed in Appendix 5-1 in this text for the readers' convenience.

When using R or U or k of a building material, it is important to pay attention to the thickness, L. In some literature, the thermal resistance may be expressed in terms of resistance values per unit thickness, or the resistance values as it is at the thickness of the product. For example, the cellular polyurethane has an RSI value of 43.38 $\left(\frac{℃ \cdot m}{W}\right)$ per meter thickness. Then for a cellular polyurethane plate with a 5 cm thickness, the RSI value will be

$$RSI\Big|_{5\text{ cm}} = 43.38\left(\frac{℃ \cdot \text{m}}{\text{W}}\right) \times 0.05(\text{m}) = 2.169\left(\frac{℃ \cdot \text{m}^2}{\text{W}}\right)$$

Building shelters, such as walls and ceilings, are made of several layers and each layer is made of different materials. The following two types of walls are most common in calculating heat conduction rates: multi-layer and parallel.

5.2 Heat Conduction through a Multi-Layer Wall

A multi-layer wall is a wall composed of several layers and each layer is made of homogeneous materials. This multi-layer wall can be considered as a stack of single layer walls, each layer has a thermal resistance of R_1, R_2, R_3... R_n. The total thermal resistance, R_t, is the sum of each single layer,

$$R_t = \sum_{i=1}^{n} R_i \tag{5-6}$$

To illustrate the heat conduction rate through such a multi-layer wall, let's take a three layer wall as an example. The wall is made of three layers: dry wall (interior), fiber glass insulation and wood sheathing (exterior), with thermal resistance values of R_1, R_2 and R_3, respectively (Figure 5-3). Assume the outside and inside air temperatures are T_o and T_i, respectively, then the total thermal resistance R_t, is

$$R_t = R_1 + R_2 + R_3 \tag{5-7}$$

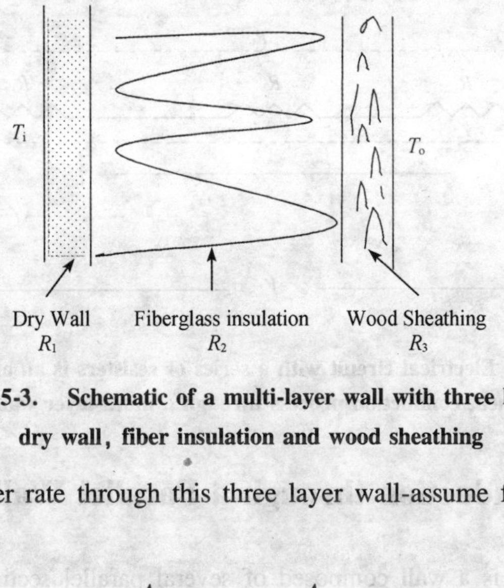

Figure 5-3. Schematic of a multi-layer wall with three layers: dry wall, fiber insulation and wood sheathing

The heat transfer rate through this three layer wall-assume from interior side to exterior side is:

$$q = -\frac{A}{R_t}(T_o - T_i) = \frac{A}{R_t}(T_i - T_o) \tag{5-8}$$

The heat transfer through a multi-layer wall can be mimicked by an electrical circuit similar to the three layer wall discussed above. Let's consider an electrical circuit with three resisters in a series, R_1, R_2 and R_3, a current I and a voltage V. As shown in Figure 5-4, the total resistance R is simply the sum of the three resisters individual resistance value.

$$R = R_1 + R_2 + R_3 \tag{5-9}$$

And the current is the voltage divided by the total resistance (Ohm's Law),

$$I = \frac{V}{R} \tag{5-10}$$

Obviously, Fourier's heat transfer law is an analogy of Ohm's Law of electricity. Comparing Equation 5-8 and Equation 5-10, they are found to be in the same format. Specifically:

- Heat transfer flux q'' ~ Electricity Current I.
- Temperature difference ΔT ~ Voltage V.
- Thermal resistance R ~ Electrical Resistance R.

Note that heat transfer rate, q, equals the heat flux, q'' multiplies area, A.

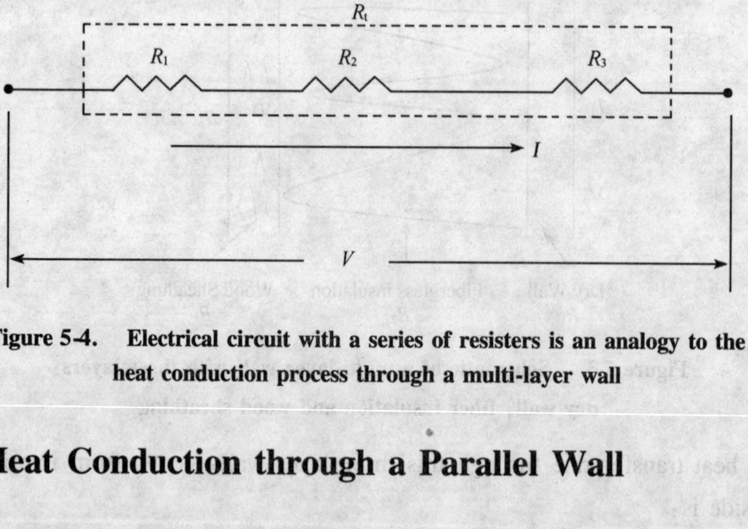

Figure 5-4. Electrical circuit with a series of resisters is an analogy to the heat conduction process through a multi-layer wall

5.3 Heat Conduction through a Parallel Wall

A parallel wall is a wall composed of several parallel sections subjected to the same temperature difference. Each section is made of homogeneous materials, but may have different areas (A_1, A_2...A_n) and different thermal resistance of R_1, R_2, $R_3 \cdots R_n$.

To calculate the heat conduction through such a parallel wall, each section of the wall can be treated as a multi-layer wall as discussed above and then add the heat transfer rates of all sections.

$$q = \sum_{i=1}^{n} \frac{A_i}{R_i} \Delta T \tag{5-11}$$

Sometimes the overall thermal resistance of the entire wall is of interest. In that case, the overall thermal resistance, R_t, can be determined as follows. To illustrate the heat conduction rate through such a parallel wall, let's take a typical wall containing a window and a door. Since the window, door and wall have a different thermal resistance and different areas (Figure 5-5), the heat transfer rate can be written as

$$q = \frac{A_1}{R_1}\Delta T + \frac{A_2}{R_2}\Delta T + \frac{A_3}{R_3}\Delta T$$
$$= \left(\frac{A_1}{R_1} + \frac{A_2}{R_2} + \frac{A_3}{R_3}\right)\Delta T \qquad (5\text{-}12)$$

Where ΔT is the temperature difference across the wall.

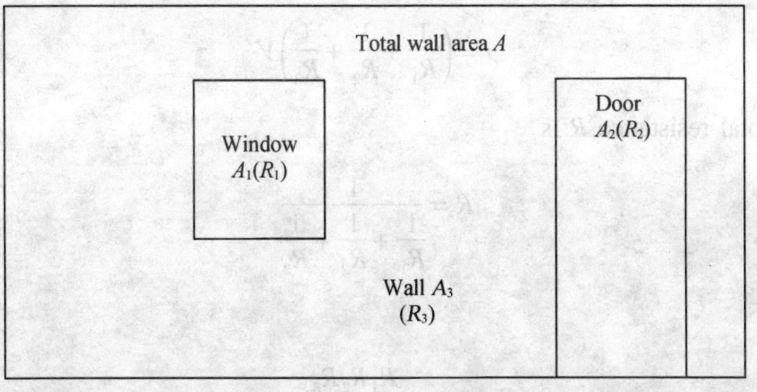

Figure 5-5. A parallel wall with a total area A, consisting of a window with an area A_1, a door with an area A_2, and a portion of solid wall with an area A_3

If the total wall area is A, of which $\frac{A_1}{A}$ is the fraction of the window area, $\frac{A_2}{A}$ is the fraction of the door area, and $\frac{A_3}{A}$ is the fraction of the solid wall area, then Equation 5-12 can be rewritten as

$$q = \left(\frac{A_1}{R_1 A} + \frac{A_2}{R_2 A} + \frac{A_3}{R_3 A}\right)A\Delta T \qquad (5\text{-}13)$$

Comparing Equation 5-13 with Equation 5-8, the equivalent total wall thermal resistance is

$$R_t = \frac{1}{\frac{A_1}{R_1 A} + \frac{A_2}{R_2 A} + \frac{A_3}{R_3 A}} \tag{5-14}$$

or

$$R_t = \frac{R_1 R_2 R_3 A}{R_2 R_3 A_1 + R_1 R_3 A_2 + R_1 R_2 A_3} \tag{5-15}$$

The heat transfer through a parallel wall can also be mimicked by an electrical circuit. Similar to the wall consisting of three sections (window, door and solid wall), let's consider an electrical circuit with three resisters in parallel: R_1, R_2 and R_3, a current I and a voltage V. As shown in Figure 5-6, the current is

$$I = \left(\frac{1}{R_1} + \frac{1}{R_2} + \frac{1}{R_3}\right) V \tag{5-16}$$

The total resistance R is

$$R = \frac{1}{\frac{1}{R_1} + \frac{1}{R_2} + \frac{1}{R_3}} \tag{5-17}$$

or

$$R = \frac{R_1 R_2 R_3}{R_1 R_2 + R_1 R_3 + R_2 R_3} \tag{5-18}$$

Example 5-1

A three layer wall of a building is composed of an interior plastic board with a *RSI* value 0.08 m^2 · ℃/W, a fiberglass insulation R20 with a *RSI* value of 3.5 m^2 · ℃/W, and an exterior layer of sheathing with medium density and a *RSI* value 0.21 m^2 · ℃/W. Assume the inside and outside temperatures are 20℃ and −20℃, respectively. Find:

1. Total insulation value of the wall, *RSI*;

2. Temperature at the interface of the interior layer and the fiberglass insulation; and

3. Temperature at the interface of the exterior layer and the fiberglass insulation.

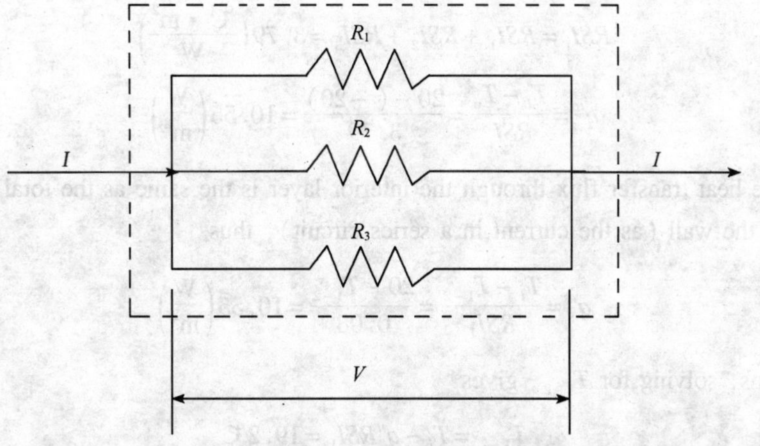

Figure 5-6. Electrical circuit with parallel resisters is an analogy to the heat conduction process through a parallel wall

Solution

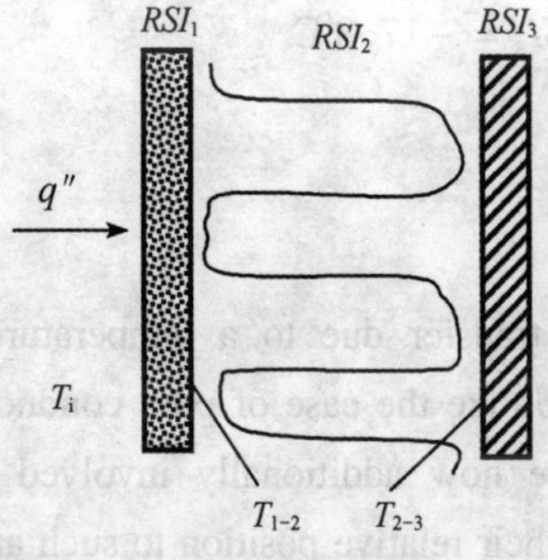

Known

RSI_1=0.08 m² • ℃/W,

RSI_2=3.5 m² • ℃/W,

RSI_3=0.21 m² • ℃/W,

T_1=20℃

T_2=−20℃

Using the electrical analogy, $I \sim q''$, $V \sim \Delta T$ and $R \sim RSI$, the wall system can be represented as

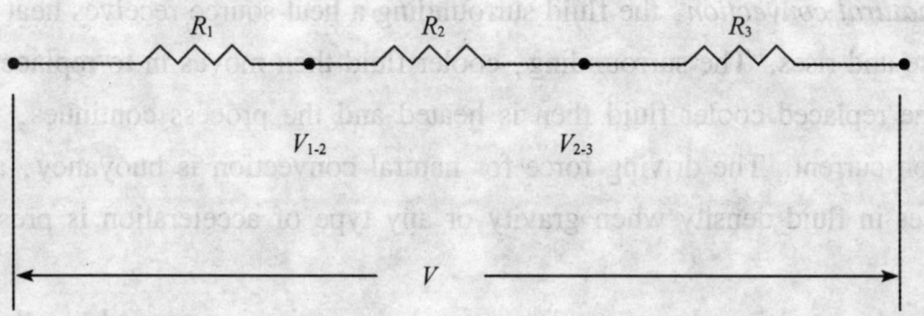

$$RSI_t = RSI_1 + RSI_2 + RSI_3 = 3.79\left(\frac{℃ \cdot m^2}{W}\right)$$

$$q'' = \frac{T_i - T_o}{RSI_t} = \frac{20 - (-20)}{3.79} = 10.55\left(\frac{W}{m^2}\right)$$

The heat transfer flux through the interior layer is the same as the total heat flux through the wall (as the current in a series circuit), thus

$$q'' = \frac{T_i - T_{1-2}}{RSI_1} = \frac{20 - T_{1-2}}{0.08} = 10.55\left(\frac{W}{m^2}\right)$$

Thus, solving for T_{1-2}, gives

$$T_{1-2} = T_i - q''RSI_1 = 19.2℃$$

Thus, solving for T_{2-3}, gives

$$q'' = \frac{T_{2-3} - T_o}{RSI_3} = \frac{T_{2-3} - (-20)}{0.21} = 10.55\left(\frac{W}{m^2}\right)$$

$$T_{2-3} = T_o + q''RSI_3 = -17.8℃$$

5.4 Convection Heat Transfer

Heat convection is a thermal energy transfer due to a temperature difference between a solid and its surrounding fluid. Unlike the case of pure conduction, which only involves solids, *currents in fluids* are now additionally involved in the heat transfer process. In solids, molecules keep their relative position to such an extent that bulk movement or flow is prohibited and, therefore, convection does not occur. While in heat convection, bulk flow movement transports thermal energy from one location to another. There are two categories of heat convection: natural and forced.

In *natural convection*, the fluid surrounding a heat source receives heat, becomes less dense and rises. The surrounding, cooler fluid then moves in to replace the rising fluid. The replaced cooler fluid then is heated and the process continues, forming a convection current. The driving force for natural convection is buoyancy, a result of differences in fluid density when gravity or any type of acceleration is present in the system.

Forced convection, by contrast to natural convection, occurs when other external driving forces provided by pumps, fans or other means to propel the fluid and create an artificially induced convection current. Forced heat convection is sometimes referred to as heat advection or simply advection. But advection is a more general process, and in heat advection, the substance being "advected" in the fluid field is simply heat (rather than mass, which is the other natural component in such situations, as mass transfer and heat transfer share generally the same equations).

In some heat transfer systems, both natural and forced convection contribute significantly to the rate of heat transfer. To calculate the rate of convection between an object and the surrounding fluids, engineers employ the convective heat transfer coefficient, h_c, or simply called heat convection coefficient. Unlike the thermal conductivity, the heat convection coefficient is not a material property. Typically, the heat convection coefficient is a much more complicated parameter than the heat conductivity in a heat conduction process. The heat convection coefficient depends upon the geometry, fluid, temperature, fluid velocity, and other characteristics of the system in which convection occurs.

The general convective heat transfer at a surface can be written in a formula similar to Fourier's Law, replacing the heat conductivity with a coefficient, h_c:

$$q = Ah_c(T_\infty - T_s) = Ah_c\Delta T \tag{5-19}$$

Where

A = area of heat convection, m^2

h_c = convective heat transfer coefficient, W/(m^2 · ℃)

T_∞ = temperature of the fluid, ℃

T_s = temperature of the surface of concern, ℃

Formulae and correlations are available in many references to calculate heat transfer coefficients for typical configurations and fluids. For example, for a liquid flowing in a straight circular pipe with a Reynolds number between 10 000 and 120 000 when the liquid's Prandtl number is between 0.7 and 120, for a location far from the pipe entrance (more than 10 pipe diameters; more than 50 diameters according to many authors) or other flow disturbances, and when the pipe surface is hydraulically smooth, the heat transfer coefficient between the bulk of the fluid and the pipe surface can be expressed as:

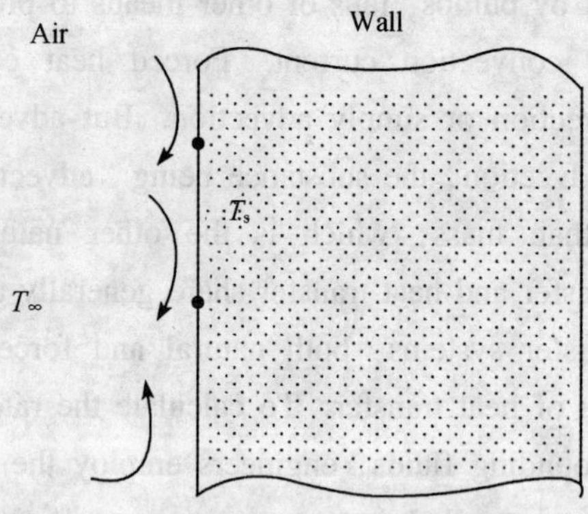

Figure 5-7. Heat transfer at a surface of a wall

$$h_c = \frac{k_l}{D_h} Nu \qquad (5\text{-}20)$$

Where

k_l = thermal conductivity of the liquid (i.e., water)

D_h = hydraulic diameter

Nu = Nusselt number = $0.023 Re^{0.8} Pr^n$

Pr = Prandtl number = $\frac{C_p \mu}{k_l}$

Re = Reynolds number = $\frac{m\mu}{D_h A}$

m = mass flow rate

μ = fluid viscosity

C_p = heat capacity of the fluid at constant pressure

A = cross-sectional area of flow

n = 0.4 for heating (wall hotter than the bulk fluid) and 0.3 for cooling (wall cooler than the bulk fluid)

The above example illustrates that although basic theory exists and various simulation models can estimate the convection coefficient, the heat transfer coefficient must be derived or found experimentally for every system analyzed.

It should be noted that convection does not occur in a perfect vacuum due to the lack of media to transmit heat. The heat convection does not occur in space where there is no atmosphere in the surroundings of the system to be analyzed. It only occurs where fluids are present.

5.5 Radiation Heat Transfer

Radiative heat transfer is a thermal energy transferred due to temperature difference between two objects by means of emitted thermal radiaton through the empty space between them. All materials radiate thermal energy in amounts determined by their temperature, where the energy is carried by photons of light in the infrared and visible portions of the electromagnetic spectrum. When temperatures are different, thermal energy is transported from surfaces of higher to surfaces of lower temperature.

All objects with a temperature above absolute zero radiate energy at a rate equal to their emissivity multiplied by the rate at which energy would radiate from them if they were a black body. No medium is necessary for radiation to occur; radiation works even in and through a perfect vacuum. The energy from the sun travels through the vacuum of space before warming the earth. Also, the only way that energy can leave the earth is by being radiated to space.

Figure 5-8. If an object has a temperature higher than zero degree absolute temperature, it is able to emit radiative heat

The radiative heat from an object to a space is related to its body surface absolute temperature, T (in kelvin); emissivity, ε; surface area, A; and the Stephan-Boltzmann constant, σ. The emissivity (or emitance) is a material property that can be found in many handbooks. Emissivity for some of the typical building materials can be found in references such as the ASHRAE Fundamentals Handbook (2005).

$$q = A\varepsilon\sigma T^4 \tag{5-21}$$

Where

$$\sigma = 5.669\ 744 \times 10^{-8} \frac{\mathrm{W}}{\mathrm{m}^2 \cdot \mathrm{K}^4} \tag{5-22}$$

ε = Emissivity, dimensionless between 0 to1. For black body, $\varepsilon = 1$

T = Absolute temperature, K

The emissivity of a material is the ratio of energy radiated by a particular material to energy radiated by a black body at the same temperature. It is a measure of a material's ability to radiate absorbed energy. Emissivity is dimensionless. When an object is a receiver of radiative heat, the absorptivity (absorptance) of the object is the opposite of its emissivity, but in the same value.

It is worth mentioning that emissivity is different from reflectivity. Both *reflectivity* and *emissivity* of all bodies is wavelength dependent. The temperature determines the wavelength distribution of the electromagnetic radiation. For any body the reflectivity depends on the wavelength distribution of incoming electromagnetic radiation and, therefore, the temperature of the source of the radiation. The emissivity depends on the wave length distribution and, therefore, the temperature of the body itself. For example, fresh snow- which is highly reflective to visible light (reflectivity about 0.90)- appears white due to reflecting sunlight with a peak energy wavelength of about 0.5 micrometers. Its emissivity, however, at a temperature of about −5℃ and a peak energy wavelength of about 12 micrometers, is 0.99.

Example 5-2

Find the radiant heat from a black iron wood stove, which has $\varepsilon = 0.9$, surface area 0.8 m^2 and surface temperature 400℃.

$$q = 0.8(\mathrm{m}^2) \times 0.9 \times 5.669\ 7 \times 10^{-8}\left(\frac{\mathrm{W}}{\mathrm{m}^2 \cdot \mathrm{K}^4}\right) \times (673\ \mathrm{K})^4 = 8\ 374\ (\mathrm{W})$$

5.6 Radiation Heat Exchange

When the temperature of Object 1 is higher than Object 2, the net radiative *heat flux* from Object 1 to Object 2 is

$$q''_{1-2} = \varepsilon_1 \sigma (T_1^4 - T_2^4) \tag{5-23}$$

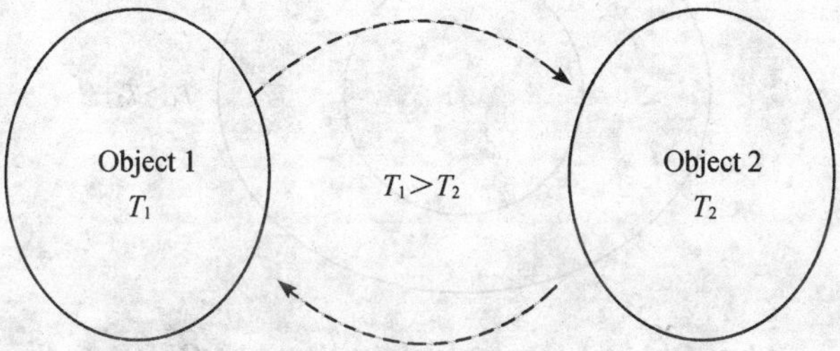

Figure 5-9. When Object 1 has a higher temperature than Object 2, a net radiative heat transfer occurs from Object 1 to Object 2

For example, all heat on the earth is from the radiative energy from the Sun. The heat flux from the Sun for the Earth is 1.39 kW/m^2, varying slightly due to the climatic conditions.

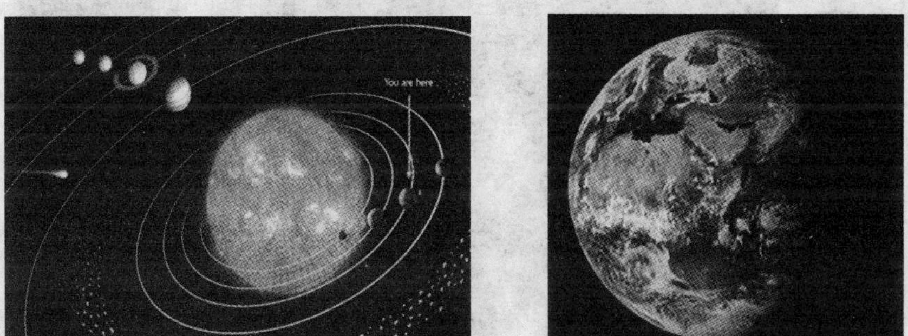

Figure 5-10. Radiative heat from the Sun to the Earth

The calculation of radiative heat exchange between two objects involves determining the angle factors between two objects. The angle factor determines the efficiency of radiative heat to the receiving surface. This text will not cover the angle factor calculations.

However, there is one case that is rather simple but useful in building environments. When Object 1 is completely enclosed by Object 2, as shown in Figure 5-11, the net radiative heat received by Object 2 from Object 1 is:

$$q_{1-2} = A_1 \varepsilon_1 \sigma (T_1^4 - T_2^4) \tag{5-24}$$

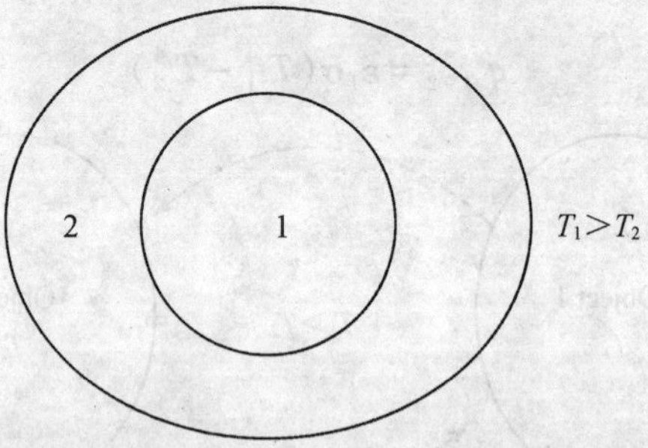

Figure 5-11. When Object 1 is completely enclosed by Object 2, the radiant heat from Object 1 is completely absorbed by Object 2

An example of this special case is a stove in a room, where the stove (Object 1) is completely enclosed by the room (Object 2).

Figure 5-12. A stove is completely enclosed by a room, thus all the radiant heat of the stove is absorbed by the room

Example 5-3

From the last example (Example 5-2, the stove), the stove surface temperature is 400℃. Assume all interior wall surfaces of the room are 25℃. How much radiant heat is emitted to the room from the stove?

Solution

Since the stove is contained in the room, from Equation 5-24,

$$q_{1-2} = A_1 \varepsilon_1 \sigma (T_1^4 - T_2^4)$$

$$q_{\text{stove}\rightarrow\text{room}} = 0.8\ (\text{m}^2) \times 0.9 \times 5.669\,7 \times 10^{-8} \times (673^4 - 298^4)\left(\frac{\text{W}}{\text{m}^2 \cdot \text{K}^4}\right)$$
$$= 8\,374\ (\text{W}) - 322\ (\text{W})$$
$$= 8\,052\ (\text{W})$$

The net heat received by the room from the stove is slightly lower than the radiant heat from the stove as shown in Example 5-2, which is 8 374 W.

5.7 Problems

1. Define the following variables and their dimension: *R*-value, *U*, *k*, *h* (here *h* is the convective heat transfer coefficient, not enthalpy), ε, σ.

2. Consider a wall made of three layers. The outer layer is 10 cm of concrete block (rectangular core, sand and gravel aggregate, 2 cores), the inner layer is the same and the center layer is made of 5 cm of expanded polystyrene molded beads with a density of 16 kg/m^3. Outside and inside temperatures are 11 and 25℃, respectively, find:

(**Hint**: use Appendices to find needed thermal properties.)

a. Total thermal resistance of the wall, *RSI*.

b. If the wall contains 10% total wall area of windows that are made of double insulating glass, 6 mm air space and 3 mm glass, and 5% total wall area of wood doors, door type is no storm doors, flush with solid core, 44 mm thick. What is the overall *RSI* of wall now?

c. Temperatures at the interfaces between the inner and center layers and between the outer and center layers.

3. A livestock building is heated with hot water pipes. The pipes are painted

black to increase the emittance to 0.9. The total pipe surface area is 1.5 m^2 and the average pipe surface temperature is maintained at 70℃. The room air temperature is 22℃ and the temperature of interior surfaces is approximately the same as the room air temperature. It is estimated that the average convective heat transfer coefficient, h, is 8 $W/m^2 \cdot ℃$. Find:

a. How much heat is gained by the building through heat convection?

b. How much heat is gained by the building through radiation?

4. What is the total heating load (in Watt) for a lecture auditorium with a room temperature 24℃ holding 200 students? The average body surface area of each student is 1.8 m^2 and they are seated and reading. How many kg of moisture will be produced in one hour if 50% of the total heat is latent heat? (Use ASHRAE data for calculation.)

Module 6: Heat Transfer through a Building Envelope

In this Module, we will discuss the heat transfer rate through typical building envelopes, especially light frame buildings, which are most common in the livestock, greenhouse and residential house sectors.

A typical light-frame structure (Figure 6-1) consists of walls, ceiling, floor, and an exterior perimeter. The exterior perimeter is the intersecting line of exterior walls with the ground, which could be a major source of heat loss during cold weather. In Figure 6-1, the ceiling consists of two layers, an interior dry-board or a plywood board, and a layer of insulation (fiberglass or blown-in insulations). The wall consists of three layers, an interior board, a layer of insulation and an exterior sheathing. Walls and ceilings are wrapped with a plastic vapor barrier. However, the barrier is so thin that it is negligible from the heat transfer point-of-view. The floor is made of a single layer of 10 cm (4 inch) thick concrete without insulation. The foundation was insulated with a 5 cm (2 inch) polyethylene rigid insulation.

The following discussions are heat transfer through four primary components of a light-frame building envelope: wall, ceiling, floor, and exterior parameter. Heat transfer through all these components, within the context of this course, can be simplified as one-dimensional, steady-state heat transfer. Each layer of material in the envelope is homogeneous. Therefore, the heat transfer can be directly derived from Fourier's Law.

6.1 Heat Transfer through a Wall

The heat transfer through a wall, q_w, in Watt, is

$$q_w = \frac{A}{RSI_w}(T_i - T_o) \tag{6-1}$$

Where

A = area of the wall, m^2

RSI_w = thermal resistance of the wall, $\frac{℃ \cdot m^2}{W}$

T_i, T_o = temperatures of inside and outside, respectively, ℃

Note: The RSI_w must be determined properly, depending on whether the wall is a multi-layer, or a parallel, or a combination of the two. Thermal resistance values for typical light-frame building materials are listed in Table 6-1.

6.2 Heat Transfer through a Ceiling

The heat transfer through a ceiling, q_c, in Watt, is

$$q_c = \frac{A}{RSI_c}(T_i - T_{attic}) \tag{6-2}$$

Where

A = area of the ceiling, m^2

RSI_c = thermal resistance of the wall, $\frac{℃ \cdot m^2}{W}$

T_i = temperatures of inside, ℃

T_{attic} = temperatures of attic, ℃

For the purpose of estimation of heat transfer through buildings on a time-weighted average, we assume that the attic temperature is

$$T_{attic} = T_o + \delta T \tag{6-3}$$

Where $\delta T = 2℃$ for winter and $3℃$ for summer.

It should be pointed out that the temperature of the attic, as shown in the structure of Figure 6-1, could be substantially different from the outside temperature. For non-insulated roofs in typical light-frame structures, temperature in an attic could be tens of degrees higher than the outside temperature during a sunny summer day. Therefore, residential houses now have vents on the roof and some have fans to ventilate the attic airspace.

6.3 Heat Transfer through a Floor

If a floor is between floor levels , then there is usually no heat transfer because

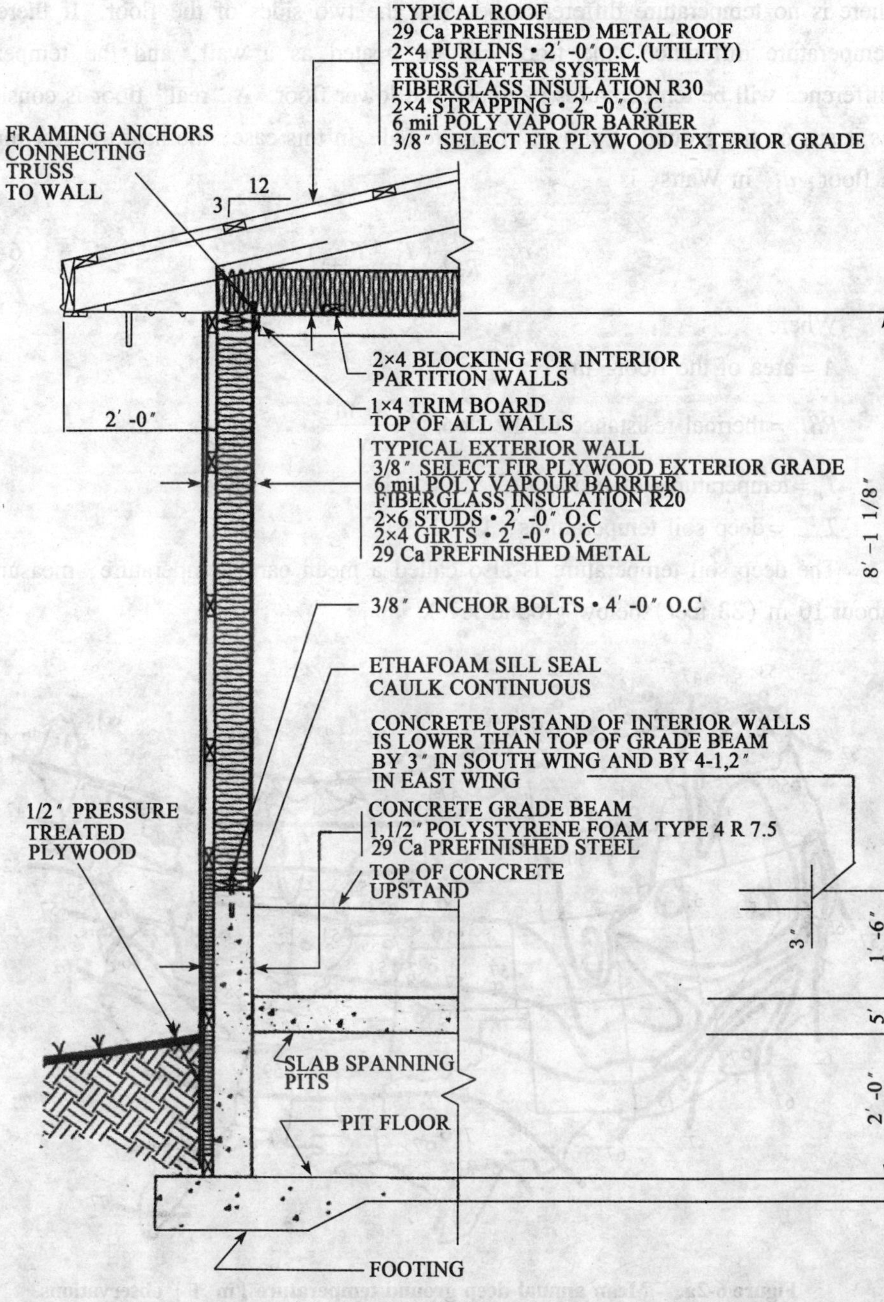

Figure 6-1. Typical structure and insulation levels of a light-frame structure

there is no temperature difference between the two sides of the floor. If there is a temperature difference, the floor can be treated as a wall, and the temperature difference will be temperatures of upper and lower floor. A "real" floor is considered as it sits on the ground as shown in Figure 6-1. In this case, the heat transfer through a floor, q_f, in Watts, is

$$q_f = \frac{A}{RSI_f}(T_i - T_{deep}) \tag{6-4}$$

Where

A = area of the floor, m^2

RSI_f = thermal resistance of the floor, $\frac{℃ \cdot m^2}{W}$

T_i = temperatures of inside, ℃

T_{deep} = deep soil temperatures, ℃

The deep soil temperature is also called a mean earth temperature, measured at about 10 m (33 feet) below ground level.

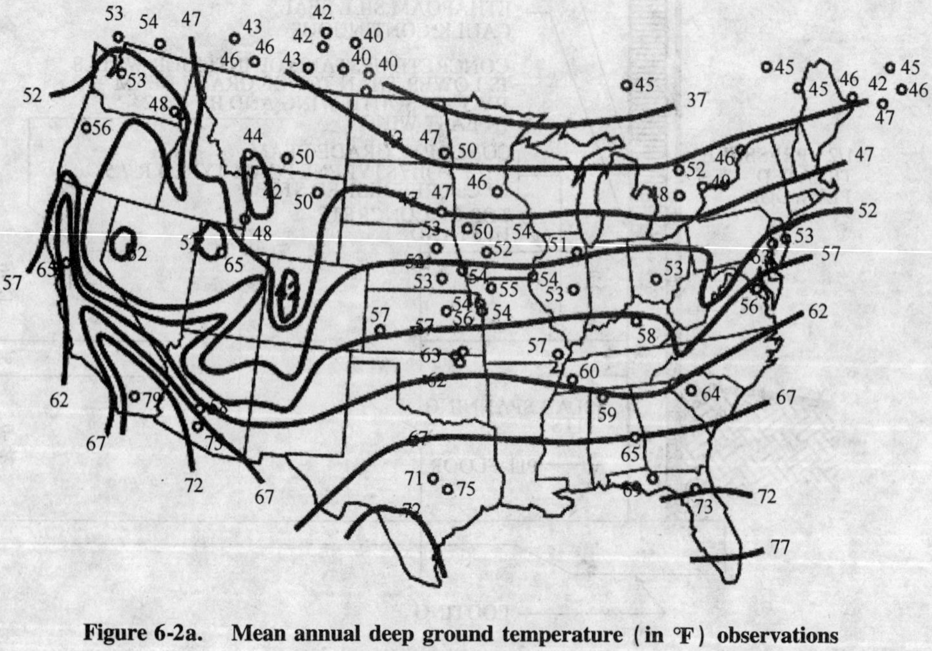

Figure 6-2a. Mean annual deep ground temperature (in ℉) observations at individual stations in the U. S

At soil depths greater than 10 meters below the surface, the soil temperature is relatively constant and corresponds roughly to the water temperature measured in groundwater wells 30 to 50 feet deep. This is referred to as the "mean earth temperature" or "deep ground temperature" in this text. Figure 6-2a shows the mean earth temperature contours across the United States. In Central Illinois, the mean earth temperature is about 54°F (12°C).

Figure 6-2b shows the average temperature of provincial capital cities in China in winter (from November to February) of 2000 to 2007.

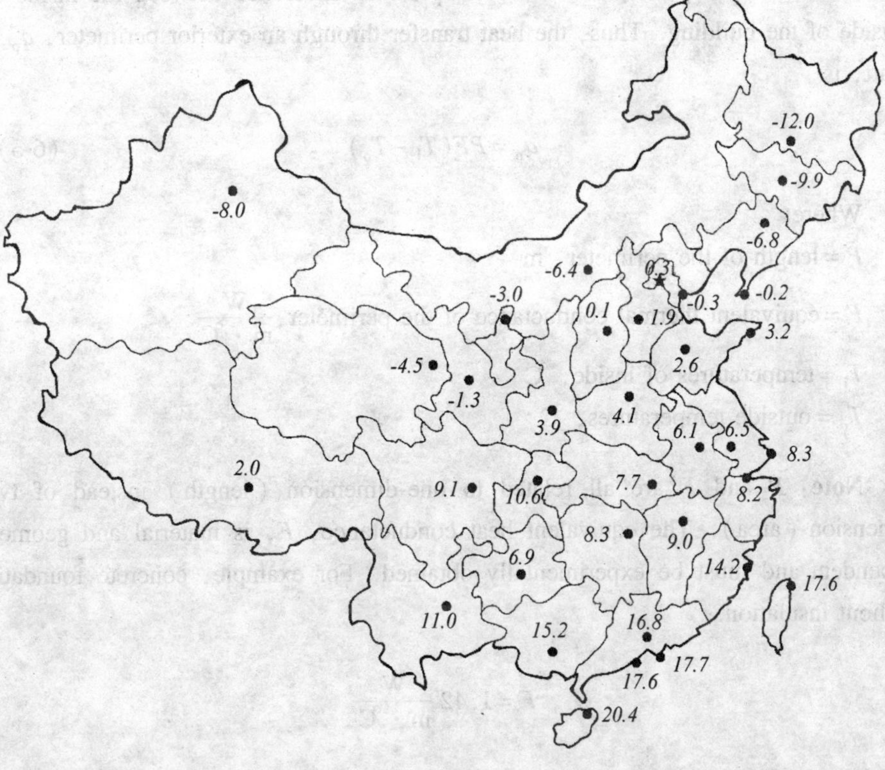

Figure 6-2b. Average temperature (in ℃) of November to February between 2000 and 2007 observations at the provincial capital cities in China (Sources: China Meteorological Yearbook, 2000 to 2007, Beijing, Meteorological Press)

6.4 Heat Transfer through Exterior Perimeter

A building perimeter refers to the intersection of the outside wall with the ground. The heat transfer in the vicinity of the building perimeter is actually very complex, involving three-dimensional heat transfer. For many light structure buildings, the heat transfer through the perimeter, on a time average, can be estimated as proportional to the length of the perimeter, an equivalent heat conductance of the perimeter and the temperature difference between the inside and outside of the building. Thus, the heat transfer through an exterior perimeter, q_p, in Watt, is

$$q_p = PF(T_i - T_o) \qquad (6\text{-}5)$$

Where

P = length of the perimeter, m

F = equivalent thermal conductance of the perimeter, $\frac{W}{m \cdot ℃}$

T_i = temperatures of inside, ℃

T_o = outside temperatures, ℃

Note: P and F are all related to one-dimension (length) instead of two-dimension (area). The equivalent heat conductance, F, is material and geometry dependent and must be experimentally obtained. For example, concrete foundation without insulation.

$$F = 1.42 \frac{W}{m \cdot ℃}$$

and concrete with 5 cm (2 inch) polyethylene rigid insulation

$$F = 0.4 \frac{W}{m \cdot ℃}$$

Table 6-1. Insulation (*RSI*) and perimeter equivalent heat conductance (*F*) values of common building materials

Material	*RSI* (m² · ℃/W)	*F* [W/(m · ℃)]
Walls with R20 insulation	3.94	
Ceiling with R30 insulation	5.48	
Floor with solid concrete	1.76	
150 mm (6 inch) R20 fiberglass	3.5	
200 mm (8 inch) R30 fiberglass	5.25	
9.5 mm (3/8 inch) plywood	0.08	
50 mm (2 inch) polyethylene rigid insulation	1.24	
150 mm concrete wall	0.08	
Concrete floor to soil below	1.76	
exterior sheet metal	0.0	
Polyethylene film vapor barrier	0.0	
Air space between layers (<2 mm)	0.2	
Inside surface air film	0.12	
Outside surface air film	0.03	
Concrete without insulation		1.42
Concrete with 25 mm polyethylene rigid insulation and 300 mm below exterior grade		0.85
Concrete with 50 mm polyethylene rigid insulation and 300 mm below exterior grade		0.42

Sources: Agriculture Canada, 1983, Publication 1601/E; ASHRAE Fundamentals, 2005.

6.5 Case Study 1: Improve energy efficiency for a building

6.5.1 House Structure Specifications (Total floor area: 1 540 ft²)

1. Floor to ceiling height: 9 ft; crawl space has an average temperature of 5℃.
2. Windows: 3 ft × 5 ft ($W \times H$) per unit, except living room window is 3 × 6

($W \times H$). All with double glass.

3. Doors: 3 ft × 7 ft. ($W \times H$) per unit.

4. Covered porch and garage are considered as outside.

5. Average outside $T_o = -6℃$, $\Phi_o = 80\%$, room $T_i = 22℃$, $\Phi_i = 30\%$.

The upgraded structure includes additional insulation and air tightness. See Table 6-2 below.

Table 6-2. Insulation and air exchange for typical and upgraded structures

Components	Typical	Upgraded
Wall, R15, total *RSI* (m² · ℃/W)	3.07	3.07
Ceiling with a total *RSI* (m² · ℃/W)	3.65	7.2
Floor, particle board (with a crawl space at 5℃) (m² · ℃/W)	0.776 (1″ PE insulation)	3.05 (insulation)
Perimeter insulation value, *F* [W/(m · ℃)]	1.42	0.42
Windows	0.5	0.5
Doors	0.5	0.5
Air leakage (air change per hour)	1	0.25

6.5.2 Find

1. What is the heating bill for the month of January, if heat (natural gas) is 3 cents per kW · h equivalent?

2. What is the percentage of energy savings for an upgraded house compared with a typical house?

3. Food for thought (*Optional*): How long before there is a return on the investment in upgrades, if the added insulation cost is \$1.5/m^2 for the ceiling, \$15/m^2 for floor and \$5/m for the perimeter?

6.5.3 Suggestion

In calculating heat loss in a typical January, split the class into an even number of teams. Half of the teams work on the calculations for a *typical* house and the other half work on the *Upgraded* house. First, each team works together to calculate perimeters, reasonable estimations are acceptable. Then among the team members, divide different tasks such as the following:

1. Heat loss through walls, windows and doors.
2. Heat loss through ceiling, floor and perimeters.
3. Heat loss through air leakage.

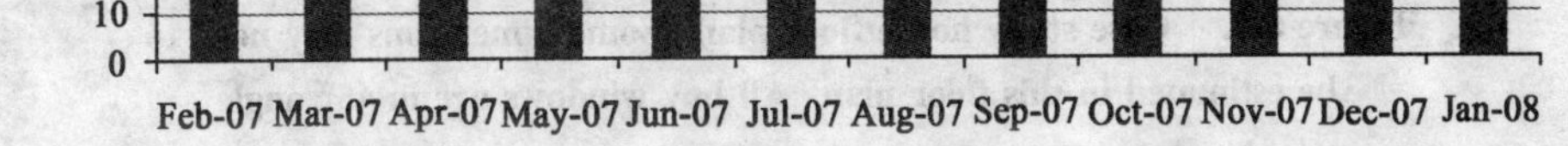

Figure 6-3 Illinois temperature graph

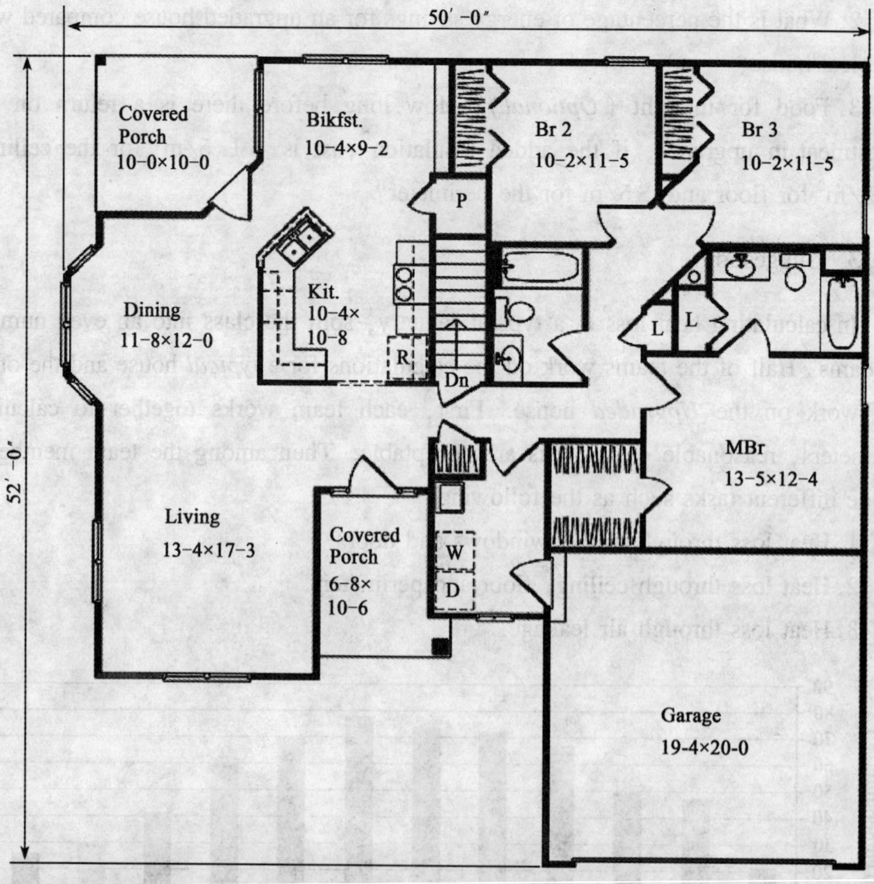

Figure 6-4. Case study house floor plan. Some dimensions may need to be estimated in this floor plan. All bay windows are in 45° angle

An Example of Energy Calculation for the Case Study

Case Study-Building Heat Transfer and Heating Cost						
Component		Typical		Upgraded		Cost for Upgrade
		RSI	H-L, W	RSI	H-L, W	
Solid wall (m^2)	143.10	3.07	1 305	3.07	1 305	
Windows (m^2)	22.20	0.50	1 243	0.50	1 243	
Doors (m^2)	5.82	0.50	326	0.50	326	
Total Wall (m^2)	171.10		2 874		2 874	

For a free Energy Design Guide download, see next page and visit http://www.ashrae.org/publications/page/1604

Continued

Case Study-Building Heat Transfer and Heating Cost

Component		Typical		Upgraded		Cost for Upgrade
		RSI	H-L, W	RSI	H-L, W	
Ceiling (m^2)	142.30	3.65	1 170	7.20	593	$ 213
Floor (m^2)	142.30	0.78	3 117	3.05	793	$ 2 135
Exterior Perimeter (m)	57.00	1.42	2 266	0.42	670	$ 285
Air Leakage (ACH)		1.00	5 126	0.25	1 282	
Grand Heat Loss			14 554		6 212	
Capital for upgrade						$ 2 633
Heating cost (3 cents/kW · h)/month			$ 314.37		$ 134.18	
Annual est cost (50% of peak)			$1 886.19		$ 805.11	
Annual saving						$ 1 081
Return for Upgrade: (Year)						2.44
Volume = 389.4 m^3						
Room air v = 0.844 m^3/kg						
$0.03 per kW						
Inside temp. (℃)	22					
Outside temp. (℃)	−6					
Crawl space temp. (℃)	5					

Additional Information for Case Study 1:

Advanced Energy Design Guides Free Download

The *Advanced Energy Design Guide* series provides a sensible approach to easily achieve advanced levels of energy savings without having to resort to detailed calculations or analysis. The four-color guides offer contractors and designers the tools, including recommendations for practical products and off-the-shelf technology, needed for achieving a 30%

energy savings compared to buildings that meet the minimum requirements of ANSI/ ASHRAE/IESNA Standard 90.1-1999. The energy savings target of 30% is the first step in the process toward achieving a net-zero energy building, which is defined as a building that, on an annual basis, draws from outside resources equal or less energy than it provides using on-site renewable energy sources.

These guides have been developed through the collaboration of ASHRAE, the American Institute of Architects (AIA), the Illuminating Engineering Society of North America (IESNA), and the U. S. Green Building Council (USGBC), with support from the Department of Energy (DOE), to help meet all of an owner's energy performance requirements. In an effort to promote building energy efficiency, ASHRAE and its partners have made these guides available for download (PDF) at no charge.

For these categories of buildings: Click here to begin your free download

K-12 School Buildings

For elementary, middle, and high school buildings, which have a wide variety of heating and air-conditioning requirements. Options for daylighting, an important component in schools, are included. I-P units.

Small Retail Buildings

For retail buildings up to 20 000 ft^2, the bulk of retail space in the U. S. Addresses typical uses: retail (other than shopping malls); strip shopping centers; automobile dealers; building material, garden supply, and hardware stores; department stores; drugstores; equipment and home furnishing stores; liquor stores; and wholesale goods (except food). I-P units.

Small Office Buildings

For office buildings up to 20 000 ft^2 the bulk of office space in the U. S. ; and provides benefits and savings for the building owner while maintaining quality and functionality of the office space. Awards: USGBC 2005 Leadership Award; Stars of Energy Efficiency Award, Honorable Mention (Alliance to Save Energy); and Best Sustainable Practice Award, Honorable Mention (SBIC). I-P units.

6.6 Case Study 2: Improve energy efficiency in a watch assembly plant of Beijing, China

6.6.1 Given

Building specifications are as follows:

1. Ceiling: 8 m in length, 5 m in width and 3 m in height (floor to ceiling), reinforced concrete structure with 25 mm aerated concrete insulation, a total *RSI* is 0.86 $m^2 \cdot ℃/W$;

2. Wall: 78 m^2, brick with a *RSI* of 0.645 $m^2 \cdot ℃/W$;

3. Windows: 16 m^2, double glass, *RSI* is 0.5 $m^2 \cdot ℃/W$;

4. Door: 5 m^2, woods with a *RSI* of 0.53 $m^2 \cdot ℃/W$;

5. Floor: 40 m^2, solid concrete with a *RSI* of 1.76 $m^2 \cdot ℃/W$; In Beijing, the mean earth temperature is about 13℃;

6. Air exchange rate per hour (ACH) = 0.5;

7. Average outside $T_o = -3℃$ (Figure 6-5), $\Phi_o = 85\%$, room $T_i = 20℃$, $\Phi_i = 30\%$.

The same building using an upgraded structure includes additional insulation in the ceiling and wall. The upgraded structure insulation levels are: wall with a *RSI* 3.94 $m^2 \cdot ℃/W$; and the ceiling with a *RSI* 5.48 $m^2 \cdot ℃/W$. The air exchange per hour is reduced to 0.25.

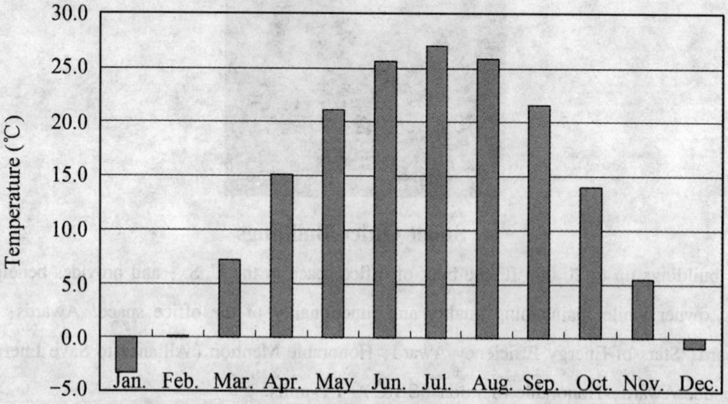

Figure 6-5. Monthly mean temperature of Beijing (2000 to 2007)

6.6.2 Find

1. Calculate the heat loss through the ceiling, walls, windows, doors and floor;

2. What is the heating bill in January, if heat (air conditioner by electricity) is 0.75 Yuan RMB per kW · h equivalent?

3. After upgraded, what is the percentage of energy savings?

6.6.3 Solution

The heat losses through the building ceiling, wall, window, door and floor and the heating cost before and after upgraded are shown in Table 6-3.

The heat loss and heating bill for the typical building in January are 6 069 W and 3 277 Yuan RMB, respectively, whereas after ungraded, the percentage of energy saving reaches 67%.

Table 6-3. Heat losses and heating cost for the typical and upgraded buildings

Components	Typical				Upgraded		
	Area (m^2)	*RSI* ($m^2 \cdot ℃/W$)	Heat loss (W)	Heating cost (Yuan)	*RSI* ($m^2 \cdot ℃/W$)	Heat loss (W)	Heating cost (Yuan)
Ceiling	40	0.86	1 070	578	5.48	168	91
Solid wall	78	0.645	2 781	1 502	3.94	455	246
Windows	16	0.5	736	397	0.5	736	397
Doors	5	0.53	217	117	0.53	217	117
Floor	40	1.76	159	86	1.76	159	86
Air exchange			1 106	597		276	149
Total		6 069	3 277			2 011	1 086

6.7 Laboratory Exercise: Heat transfer through building shelters

6.7.1 Objective

Experimentally determine the *R*-values of three different materials and compare with literature values.

6.7.2 Summary

Each model house will have an inside and an outside thermocouple attached to the wall and to the roof (total of 4) to measure the change in temperature across the material produced by heating the house with a 25 W light bulb. The conductive heat transfer equation will then be used to calculate the *R*-value of each material.

6.7.3 Equipment

1. Model houses, each built from sheet metal, plywood and rigid glass fiber insulation board (published *R*-value: 4 h – ft^2 – ℉/Btu-in)

2. 3 ×25 W light bulbs

3. 12 "T-type" thermocouples

4. IO Tech Personal DAQ 56 data acquisition unit

5. DAQ View software

6. IRISYS Multi-purpose thermal imager (IRI 4010)

6.7.4 Procedure

Note: Read the "Before you start" section on the next page before beginning the lab.

1. Measure the total surface area (interior!) of each building and the thickness of each material.

2. Connect thermocouples to high/low outlets for individual channels on the DAQ.

3. Start the DAQ View software. Each row of the spreadsheet represents a channel on the DAQ. Label each channel (row) with an appropriate name. Change the "range" column to "T-type" and the "on" column to "on".

4. Make sure the house is securely fitted over its base and then turn the light bulb on.

5. Let the DAQ collect temperature data for each thermocouple until there is no change in recorded temperature (steady state).

6. Use the Thermal Imager to record the outside temperature of each house (both the Wall and the Top). Pictures should be taken every 5 minutes until the thermocouple reading reaches the steady state. (The outside temperature can be read by using the measurement cursor to pick the position on the image.)

7. Save the data of DAQ into a spreadsheet and share with whole class.

6.7.5 Find

1. Generate the figures of relationships between temperature and time (Temperature vs. Time) for each house based on the data collected by the DaqView software.

2. Total R-value of each material calculated from the wall data and from the roof data.

3. R-value per thickness of each material.

4. Time to reach steady state.

5. Literature values for each material.

6.7.6 Discussion Questions

1. Are there any differences between the results from IRI 4010 and from the Thermocouple? What reasons do you think can lead to these differences?

2. Are the calculated R-values different when the temperature difference is measured on the roof as opposed to at the wall? Why or why not? What implication does this have?

3. Are the calculated R-values different from those listed in the literature? Why or why not? What implication does this have?

4. What does steady state have to do with it? Why might time to reach steady state influence building design?

6.7.7 Before you start, pay attention to the following

1. Thermocouple is connected with red (–) wire to analog Lows (1L, 2L, etc.) and blue (+) wire to analog Highs (1H, 2H, etc.).

2. Set up your Data Destination before you do Channel Configuration (both in the pull-down menu of View). Make sure your data file is in ASCII format, it can then be opened by Excel.

3. Change the "on" columns, which you do not connect with thermocouple, to "off".

4. Remember to click the "Arm Trigger for Disk Recording" button from the Toolbar of DaqView. Only clicking the "Enable Readings Column" from the Channel Configuration Window Toolbar will NOT record your data on the disk.

5. Remember to record the time to reach steady state.

6. The IRI 4010 uses a Class 2 laser. **Under no circumstances should anyone look directly into the laser beam or the laser beam exit aperture, irreversible damage to the eye may occur.** The laser should not be operated when there are personnel in the imager's field of view.

7. You should try to get the average temperature or temperate range of the outside Wall and the outside Top by using the cursor to pick many positions on each of the house images.

8. Save the picture taken by the IRI 4010 to the SD card and write down its name. The TA can then send the files back to you based on your requests.

Module 7: Heat and Moisture Production within Buildings

Heat and moisture (water vapor) production within a building plays an important role in the heat and mass balance of a building airspace. There are several sources of heat and moisture production within a building. Equipment, lights, occupants (people/animals), and even management activities all contribute heat and/or moisture to the room airspace. This module discusses the heat and moisture production from different courses within a building airspace.

7.1 Sensible Heat from Lights

Even with the most efficient light, most of the electricity input is converted into heat. For example, with an incandescent light, more than 95% of its energy becomes heat that is dissipated into the room. Therefore, for building heating/cooling load calculations, the sensible heat production from the light can be estimated by the power of the light, i. e., we consider a 60 W light bulb gives 60 W of sensible heat to the room.

7.2 Sensible Heat from Equipment

Similar to lights, for electrical and electronic equipment one can use the power rating of the equipment to approximate the sensible heat output to the room. For example, a computer rated at 40 W power approximately releases 40 W sensible heat to the ambient. Some equipment, such as a refrigerator, may be rather deceiving as it appears to absorb and give off heat at the same time. In fact, the refrigerator gives off approximately the same amount of sensible heat as its electricity input.

For mechanical equipment, the sensible heat production can be estimated by its power rating minus the product of the efficiency and the power rating of the equipment.

$$q_e = \text{Power rating} \times (1 - \xi_e) \tag{7-1}$$

Where

ξ_e = efficiency of the equipment.

For example, a fan motor of 200 W power rating with 30% efficiency will have a sensible heat production rate to the room, when in operation.

$$q_e = 200\ \text{W} \times (1 - 0.3) = 140\ \text{W}$$

7.3 Heat and Moisture from Management Activities

Building management activities can substantially contribute to the heat and moisture in the air. For example, power washing a barn or mopping the floor can add moisture and, hence, increase the moisture content to the building airspace. However, it is difficult to estimate this term accurately and very often reasonable assumption has to be used when data is not available.

7.4 Latent Heat and Moisture Relationship

In many cases, one needs to know the latent heat content change in the air at a given temperature when only moisture production data is available, or vise versa. From the relationship of latent heat with moisture content and temperature,

$$h_w = W\,(2\,501 + 1.805 \cdot T)$$

By multiplying a mass flow rate, m in kg/s of moist air, at the both sides of above equation, the latent heat production rate, q_w, and the moisture production rate, m_w, can be obtained as

$$q_w = m_w (2\,501 + 1.805 \cdot T) \tag{7-2}$$

Where

m_w = moisture production rate, kg/s

T = temperature, ℃

7.5 Heat and Moisture Production of Occupants

The total heat production rate of humans at various activity levels are listed in Table 7-1. A unit used to express the metabolic rate per unit DuBois area is the ***met***, defined as the metabolic rate of a sedentary person (seated/quiet): 1 met = 58.1 W/m^2 = 50 kcal/(h · m^2). A normal, healthy man at age 20 has a maximum capacity of approximately M_{act} = 12 met, which drops to 7 met at age 70. Maximum rates for women are about 30% lower. Long-distance runners and trained athletes have maximum rates as high as 20 met. An average 35-year-old who does not exercise has a maximum rate of about 10 met, and activities with M_{act} > 5 met are likely to prove exhausting.

The latent heat is related to both the activity level and environmental conditions. Some typical values of latent heat production of humans are listed in Table 7-2. Heat production can be substantially different from source to source. It is suggested that the total metabolic energy (total heat) production in Table 7-1 should be used. Data in Table 7-2 is only used as a guideline for determining the fractions of sensible and latent heat.

Livestock and poultry heat and moisture production rates are listed in Table 7-3 and Table 7-4 and the laboratory animal heat and moisture production rates are listed in Table 7-5.

It should be pointed out that the heat and moisture production rate may vary substantially from source to source for the same species of animals and poultry, depending on the measurement methods (such as calorimetry or under real housing conditions). Figure 7-1 illustrates the variation of literature data for poultry heat and moisture production rate. Two points worth mentioning are:

1. Sensible and heat production have a much larger variation than the total heat production, because latent and sensible heat change their states under certain environmental conditions.

2. The heat and moisture production for livestock, especially poultry, has increased substantially over the years because of genetic changes-as they are growing faster and tend to eat more and release more heat and moisture. For example, a broiler chicken took three months to grow to four pounds in the 1960s, but only took

about six weeks to grow to the same weight in the 1990s.

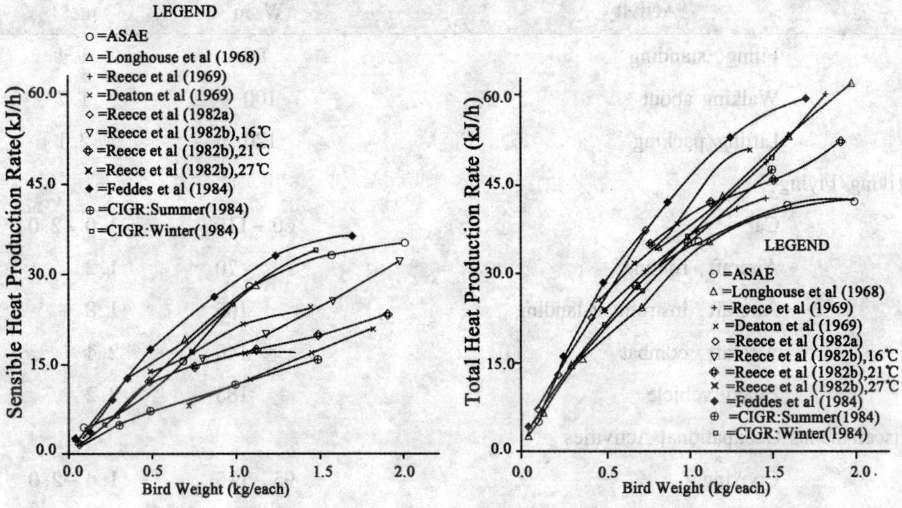

Figure 7-1. Sensible heat and total heat production of a broiler chicken from literature. The total heat has a smaller variation compared with the sensible heat

Table 7-1. Total heat production of humans during various activity levels (ASHRAE, 2005)

Activity		W/m^2	met*
Resting			
	Sleeping	40	0.7
	Reclining	45	0.8
	Seated, quiet	60	1.0
	Standing, relaxed	70	1.2
Walking (on level surface)			
	3.2 km/h (0.9 m/s)	115	2.0
	4.3 km/h (1.2 m/s)	150	2.6
	6.4 km/h (1.8 m/s)	220	3.8
Office Activities			
	Reading, seated	55	1.0
	Writing	60	1.0
	Typing	65	1.1
	Filing, seated	70	1.2

Continued

Activity	W/m²	met*
Filing, standing	80	1.4
Walking about	100	1.7
Lifting/packing	120	2.1
Driving/Flying		
Car	60 – 115	1.0 – 2.0
Aircraft, routine	70	1.2
Aircraft, instrument landing	105	1.8
Aircraft, combat	140	2.4
Heavy vehicle	185	3.2
Miscellaneous Occupational Activities		
Cooking	95 – 115	1.6 – 2.0
Housecleaning	115 – 200	2.0 – 3.4
Seated, heavy limb movement	130	2.2
Machine work		
Sawing (table saw)	105	1.8
Light (electrical industry)	115 – 140	2.0 – 2.4
Heavy	235	4.0
Handling 50 kg bags	235	4.0
Pick and shovel work	235 – 280	4.0 – 4.8
Miscellaneous Leisure Activities		
Dancing, social	140 – 255	2.4 – 4.4
Calisthenics/exercise	175 – 235	3.0 – 4.0
Tennis, singles	210 – 270	3.6 – 4.0
Basketball	290 – 440	5.0 – 7.6
Wrestling, competitive	410 – 505	7.0 – 8.7

Sources: Compiled from various sources. For additional information, see Buskirk (1960), Passmore and Durnin (1967), and Webb (1964).

* 1 met = 58.1 W/m^2.

Table 7-2. Total (metabolic) sensible and latent heat of people
(http://www.engineeringtoolbox.com/metabolic-heat-persons-d_706.html)

Degree of Activity	Typical Application	Average Metabolic rate-male adult (W)	Room Dry Bulb Temperature (℃)											
			28		27		26		24		22		20	
			Sens.	Lat.	Sens.	Lat.	Sens.	Lat.	Sens.	Lat.	Sens.	Lat.	Sens.	Lat.
Seated at rest	Cinema, theatre, school	100	50	50	55	45	60	40	67	33	72	28	79	21
Seated, very light work	Computer working	120	50	70	55	65	60	60	70	50	478	42	84	36
Office work	Hotel reception, cashier	130	50	80	56	74	60	70	70	60	78	52	86	44
Standing, walking slowly	Laboratory work	130	50	80	56	74	60	70	70	60	78	52	86	44
Walking, seated		150	53	97	58	92	64	86	76	74	84	66	90	60
Moderate work	Servant, hair dresser	160	55	105	60	100	68	92	80	80	90	70	98	62
Light bench work	Mechanical production	220	55	165	52	158	70	150	85	135	100	120	115	105
Moderate Dancing	Party	250	62	188	70	180	78	172	94	156	110	140	125	125
Fast walking	Mountain walking	300	80	220	88	212	96	204	110	190	130	170	145	155
Heavy work	Athletics	430	132	298	138	292	144	286	154	276	170	260	188	242

Table 7-3. Sensible heat, moisture and total heat production rates of livestock based on calorimetric and housing system studies (ASAE, 2000)

Livestock	Building temperature (℃)	MP[gH_2O /(kg · h)]	SHL (W/kg)	THL (W/kg)	Reference
Cattle					
Dairy cow					(Yeck and Stewart, 1959)
500 kg	-1	0.77	1.9	2.4	
500 kg	10	1.0	1.5	2.2	
500 kg	15	1.3	1.2	2.1	
500 kg	21	1.3	1.1	2.0	
500 kg	27	1.8	0.6	1.9	
					(Hellickson et al., 1974)
Beef cattle	4	2.5	1.5	2.8	
500 kg	4	2.5	1.5	2.8	
Calves					
Ayrshire male*					(Gonzalez-Jimenez and Blaxler, 1962)
39 kg(8 days)	3	0.7	2.5	2.9	
40 kg(14 days)	3	0.7	2.4	2.8	
45 kg(25 days)	3	0.7	2.6	3.0	
39 kg(8 days)	23	0.7	2.0	2.4	
40 kg(14 days)	23	0.7	1.9	2.3	
44 kg(24 days)	23	0.7	2.0	2.4	
British Friesian male*					(Holmes and Davey, 1976)
47 kg	20	—	—	2.3	
Jersey*					(Holmes and Davey, 1976)
28 kg	20	—	—	2.6	
British Friesian male*					(Webster and Gordon, 1977)
2 days to 8 wk	5	—	—	(7.9)†	
2 days to 8 wk	10	—	—	(7.3)†	
2 days to 8 wk	15	—	—	(7.3)†	

MP: moisture production, SHI: sensible heat production of livestock, THL: total heat production of livestock

Continued

Livestock	Building temperature (℃)	MP[gH_2O /(kg · h)]	SHL (W/kg)	THL (W/kg)	Reference
Cattle					
Hereford x					(Webster and Gordon, 1977)
Friesian male*					
2 days to 8 wk	5	—	—	(7.3)†	
2 days to 8 wk	10	—	—	(6.5)†	
2 days to 8 wk	15	—	—	(6.5)†	
British Friesian male*					(Webster and Gordon, 1976)
90 – 180 kg	5	(2.2)+	(6.8)‡	(8.2)‡	
100 kg	5	0.75	2.1	2.6	
90 – 180 kg	10	(4.4)+	(6.3)‡	(9.3)‡	
100 kg	10	1.4	2.0	2.9	
90 – 180 kg	15	(6.2)+	(4.6)‡	(8.7)‡	
100 kg	15	2.0	1.4	2.8	
90 – 180 kg	20	(7.1)+	(4.4)‡	(9.1)‡	
100 kg	20	2.3	1.4	2.9	
Brown Swiss Holstein					(Yeck and Stewart, 1960)
16 wk	10	2.0	2.3	3.7	
32 wk	10	1.2	1.5	2.4	
48 wk	10	1.0	1.5	2.2	
16 wk	27	3.0	1.5	3.5	
32 wk	27	2.2	1.1	2.6	
48 wk	27	1.9	1.0	2.2	
Jersey					(Yeck and Stewart, 1960)
16 wk	10	2.4	2.5	4.1	
32 wk	10	1.5	1.8	2.8	
48 wk	10	1.3	1.6	2.5	
16 wk	27	3.8	1.4	3.9	
32 wk	27	2.5	1.0	2.7	
48 wk	27	2.3	0.8	2.3	

Continued

Livestock	Building temperature (℃)	MP[gH_2O /(kg · h)]	SHL (W/kg)	THL (W/kg)	Reference
Shorthom, Brahman Santa Gertrudis					(Yeck and Stewart, 1959)
25 wk	10	1.6	1.4	2.5	
40 wk	10	1.1	1.0	1.8	
55 wk	10	1.0	1.0	1.6	
25 wk	27	2.8	0.7	2.6	
40 wk	27	2.1	0.5	1.9	
55 wk	27	1.8	0.5	1.7	
Swine					
Sow and litter (solid floor)					(Bond et al., 1959)
177 kg(0 wk)	16 – 27	1.8	1.3	2.6	
181 kg(2 wk)	16 – 27	2.4	1.7	3.3	
186 kg(4 wk)	16 – 27	2.6	1.7	3.5	
200 kg(6 wk)	16 – 27	2.7	1.7	3.5	
227 kg(8 wk)	16 – 27	2.6	2.1	3.9	
Nursery pigs*					(Ota et al., 1975)
4 – 6 kg	29	1.7	2.2	3.3	
6 – 11 kg	24	2.2	3.1	4.5	
11 – 17 kg	18	2.2	3.5	5.0	
4 kg (single)	15	—	—	6.4	(Caimle and Pullar, 1957)
	20	—	—	5.2	
	25	—	—	4.5	
	30	—	—	4.0	
6 – 8 kg (single)	15	—	—	6.0	(Caimel and Pullar, 1957)
	20	—	—	5.6	
	25	—	—	5.2	
	30	—	—	5.1	
10 – 12 kg (single)	15	—	—	5.8	
	20	—	—	5.6	
	25	—	—	5.6	
	30	—	—	5.6	

Continued

Livestock	Building temperature (℃)	MP[gH_2O /(kg · h)]	SHL (W/kg)	THL (W/kg)	Reference
Growing-finishing pigs (solid floor)					(Bond et al., 1959)
20 kg	5	2.5	4.2	5.9	
	10	2.7	4.0	5.4	
	15	3.1	3.0	5.0	
	20	3.7	2.3	4.8	
	25	4.7	1.6	4.8	
	30	6.3	0.6	4.8	
40 kg	5	1.5	3.0	4.0	
	10	1.6	2.5	3.6	
	15	1.9	2.0	3.3	
	20	2.2	1.6	3.1	
	25	2.8	1.2	3.0	
	30	3.6	0.6	3.0	
60 kg	5	1.2	2.5	3.3	
	10	1.3	2.0	2.9	
	15	1.4	1.7	2.6	
	20	1.7	1.3	2.4	
	25	2.0	1.0	2.3	
	30	2.7	0.5	2.3	
80 kg	5	1.1	2.2	2.9	
	10	1.1	1.8	2.5	
	15	1.2	1.5	2.3	
	20	1.4	1.2	2.1	
	25	1.7	0.85	2.0	
	30	2.2	0.49	1.9	
100 kg	5	0.94	2.0	2.6	
	10	1.0	1.6	2.3	
	15	1.1	1.3	2.0	
	20	1.2	1.1	1.9	
	25	1.4	0.80	1.8	
	30	1.8	0.49	1.7	

Continued

Livestock	Building temperature (℃)	MP[gH_2O /(kg · h)]	SHL (W/kg)	THL (W/kg)	Reference
Gilts, sows, and boars (solid floor)					
140 kg	5	0.79	1.8	2.3	
	10	0.79	1.5	2.0	
	15	0.84	1.2	2.0	
	20	0.93	1.0	1.6	
	25	1.1	0.77	1.5	
	30	1.3	0.54	1.4	
180 kg	5	0.64	1.7	2.1	
	10	0.63	1.4	1.8	
	15	0.65	1.2	1.6	
	20	0.70	0.97	1.4	
	25	0.80	0.80	1.3	
	30	0.96	0.63	1.3	
Sheep					
Mature*					
60 kg (maintenance diet)					
Fleece length					(Armstrong et al., 1960)
Shorn	8	0.33	2.4	2.6	
	20	0.40	1.5	1.7	
	32	0.79	0.81	1.3	
3 cm	8	0.34	1.2	1.4	
	20	0.52	0.94	1.3	
	32	1.3	0.45	1.3	
6 cm	8	0.40	1.0	1.3	
	20	1.76	0.68	1.2	
	32	1.4	0.29	1.2	
(1.3 × maintenance diet)					
Fleece length					(Alexander, 1974)
12 cm	13	1.0	0.85	1.5	
	23	1.5	0.54	1.5	
Lambs					
1 – 14 days	10	—	—	6.7	(Alexander, 1961)
(dry – no wind)	23 – 27	—	—	5.2	

* Figures are obtained from calorimetric studies. Therefore the moisture production in livestock housing systems will be increased with a corresponding decrease in sensible heat loss.

+ Figures in ( ) are g $H_2O/h \cdot kg^{0.75}$.

‡ Figures in ( ) are $W/kg^{0.75}$.

Table 7-4. Heat and moisture production of poultry (ASAE, 2000)

Poultry	Building temperature (℃)	MP[gH_2O /(kg·h)]	SHL (W/kg)	THL (W/kg)	Reference
Laying hen					
Leghorn*	8	2.1	4.5	5.8	(Ota and McNally, 1961)
	12	2.8	4.0	5.9	
	18	2.9	3.9	5.8	
	28	3.8	3.2	5.8	
Broilers*					(Longhourse, et al., 1968)
0.1 kg	29	4	12	14	
0.7 kg	25	3	7	9	
1.1 kg	19	2	7	9	
1.6 kg	19	2	6	7	
2.0 kg	19	1.5	5	6	
Broilers					
0.1 kg	29	22.0	4.5	19.5	(Reece and Lott, 1982)
0.4 kg	24	12.5	6.5	15.0	
0.7 kg	16	10.5	6.0	13.0	
	27	10.5	3.0	10.0	
1.0 kg	16	8.0	5.0	10.5	
	27	9.5	3.0	9.5	
1.5 kg	16	7.5	4.5	9.5	
	27	9.0	3.0	9.0	
2.0 kg	16	6.5	4.0	8.5	
Turkeys					
Large White* toms					
0.1 kg	35	15.6	6.1	16.8	(DeShazer et al., 1974)
0.2 kg	32	9.8	6.3	12.9	
0.4 kg	29	7.2	5.4	10.2	

Continued

Poultry	Building temperature (℃)	MP[gH_2O /(kg · h)]	SHL (W/kg)	THL (W/kg)	Reference
0.6 kg	27	4.1	5.9	8.7	
1.0 kg	24	2.4	6.3	7.9	
15 kg	25	1.7	1.1	2.2	(Shanklin et al, 1977)
hens					
8.2 kg	25	1.4	1.4	2.4	(Shanklin et al, 1977)
Wortstad white* toms					(Buffington et al, 1974)
2.2 kg	21 light	—	—	9.1	
	21 dark	—	—	6.6	
3.0 kg	21 light	—	—	8.1	
	21 dark	—	—	6.1	
3.4 kg	21 light	—	—	7.7	
	21 dark	—	—	5.8	
3.9 kg	21 light	—	—	7.1	
	21 dark	—	—	5.5	
hens					
1.6 kg	21 light	—	—	9.4	
	21 dark	—	—	6.8	
2.4 kg	21 light	—	—	8.1	
	21 dark	—	—	6.1	
2.8 kg	21 light	—	—	7.5	
	21 dark	—	—	5.7	
3.2 kg	21 light	—	—	6.9	
	21 dark	—	—	5.4	
Beltsville white* toms					(Ota and McNally, 1961)
8.9 kg	18 light	1.8	2.6	3.8	
	18 dark	1.4	2.2	3.2	

Continued

Poultry	Building temperature ℃	MP[gH_2O /(kg · h)]	SHL (W/kg)	THL (W/kg)	Reference
	25 light	2.2	1.8	3.3	
	25 dark	1.7	1.6	2.8	
hens					
4.4 kg	18 light	1.7	2.7	3.9	
	18dark	1.0	2.0	2.6	
	25 light	4.4	1.2	4.2	
	25 dark	3.0	0.7	2.8	
					(Shanklin et al., 1977)
Broad-breasted Bronze* toms					
17 kg	10	0.35	1.9	2.1	
	15	0.53	1.7	2.0	
	20	0.66	1.6	2.1	
16 kg	25	0.88	1.5	2.1	
17 kg	30	1.1	0.81	1.5	
16 kg	35	1.5	0.44	1.4	
hens					
9.8 kg	10	0.31	2.1	2.2	
9.5 kg	15	0.75	1.9	2.4	
9.5 kg	20	0.70	1.6	2.0	
9.3 kg	25	0.83	1.5	2.1	
9.1 kg	30	1.0	1.0	1.6	
8.7 kg	35	1.3	0.64	1.4	

* Data are obtained from calorimetric studies. Therefore, the moisture production in poultry housing systems will be increased with a corresponding decrease in sensible heat loss.

Table 7-5. Heat generated by laboratory animals (ASHRAE, 2005)

Animal	Mass (kg)	Heat Generation, Watts per Animal			
		Basal[a]	Normally Active[b,c]		
			Sensible	Latent	Total
Mouse	0.021	0.19	0.33	0.16	0.49
Hamster	0.118	0.70	1.18	0.58	1.76
Rat	0.281	1.36	2.28	1.12	3.40
Guinea Pig	0.41	1.79	2.99	1.47	4.46
Rabbit	2.46	6.86	11.49	5.66	17.15
Cat	3.00	7.97	13.35	6.58	19.93
Primate	5.45	12.47	20.88	10.28	31.16
Dog	10.3	20.11	30.71	16.53	47.24
Dog	22.7	36.36	67.60	36.39	103.99
Goat	36	51.39	86.08	42.40	128.48
Sheep	45	60.69	101.66	50.07	151.73
Pig	68	82.65	108.70	85.56	194.26
Chicken	1.82	5.47	3.78	6.42	10.20

[a]Based on standard metabolic rate $M = 3.5\ W^{0.75}$ watt per animal (Kleiber, 1961) or appropriate reference (W = animal mass, kg).

[b]Referenced according to availability of heat generation data. Otherwise, heat generations are calculated on basis of $ATHG = 2.5\ M$ (Gordon et al., 1976). Latent heat is assumed to be 33% of total heat and sensible heat is 67% of total heat (Besch, 1973; Woods et al., 1972).

[c]Data taken from Runkle (1964), Kleiber (1961), Besch (1973), Woods and Besch (1974), Woods et al. (1972), Bond et al. (1959), and Ota and McNally (1961).

7.6 Net Sensible Heat Transfer through Buildings

The net sensible heat transfer of a building airspace, q_n, is the sum of all heat loss or gain through and within the building airspace:

$$q_n = q_{as} + q_b + q_l + q_e + q_m + q_r \text{ (W)} \tag{7-3}$$

Where

q_{as} = sensible heat produced by animals

q_b = heat transfer through building shelter (wall, ceiling, floor and perimeters, solar radiation, etc). It is " − " if heat loss and it is " + " if heat gain

q_l = heat from light

q_e = heat from equipment

q_m = sensible heat produced from other management activities

q_r = heat gain from solar radiation

When q_n is negative from Equation 7-3, the building airspace of concern is losing heat, thus, it is also called a heating load because supplemental heat is needed to maintain a desirable temperature within the building airspace. When q_n is positive, the building airspace is gaining heat, thus, it is called a cooling load because cooling is needed to maintain the desirable temperature in the building airspace.

In reality, there is always some exchange between the forms of sensible heat and latent heat, but the total heat remains approximately constant.

7.7 Net Moisture and Pollutants Production within Buildings

Total moisture production rate within a building airspace, m_w, in mass per unit time such as $\frac{kg_{H_2O}}{h}$, is:

$$m_w = m_{wa} + m_{we} + m_{wm} \tag{7-4}$$

Where

m_{wa} = moisture production rate from animals (kg/h)

$$m_{wa} = MP \times \frac{\text{mass}}{\text{per animal}} \times \text{Number of animals} \tag{7-5}$$

MP = moisture production of animals per unit body mass $kg_{H_2O}/(kg_{body} \cdot h)$

m_{we} = moisture production rate from equipment such as a humidifier

m_{wm} = evaporation rate of water from wet surfaces due to building management practices

7.8 Carbon Dioxide Production

Airborne pollutants are numerous. One commonly used air quality indicator is carbon dioxide, which often is primarily produced by the occupants in a living airspace. Carbon dioxide is a metabolic product. Basal metabolic rate (BMR) is the amount of energy expended while at rest in a neutrally temperate environment, in the post-absorptive state (meaning that the digestive system is inactive, which requires about twelve hours of fasting in humans). BMR is measured under very restrictive circumstances when a person is awake. But at complete rest, a more common and closely related measurement, used under less strict conditions, is resting metabolic rate (RMR). BMR and RMR are measured by gas analysis through either direct or indirect calorimetry; though a rough estimation can be acquired through an equation using age, sex, height, and weight. Studies of energy metabolism using both methods provide convincing evidence for the validity of the respiratory quotient (RQ), which measures the inherent composition and utilization of carbohydrates and oxygen as they are converted to energy and carbon dioxide.

On average, every 24.6 kJ of metabolic energy or total heat produced will produce one liter carbon dioxide for human and animals under standard conditions. (This estimation is within 10% of error and can be used in most design applications). If the total heat production is in kJ per hour, then the volumetric carbon dioxide production rate, Q_{CO_2}, in m^3 of CO_2 per hour will be:

$$Q_{CO_2} = \frac{THP(\text{kJ/h})}{24\,600\left(\dfrac{\text{kJ}}{\text{m}^3 \text{ of CO}_2}\right)} \tag{7-6}$$

The CO_2 mass production rate, m_{CO_2}, is

$$m_{CO_2} = Q_{CO_2}\,\rho_{CO_2} \tag{7-7}$$

Where

m_{CO_2} = mass production of carbon dioxide (kg/h)

ρ_{CO_2} = density of the carbon dioxide = 1.83 kg/m^3, under standard room conditions

Example 7-1

How much CO_2 is produced in volume and in mass, in a room containing 30 adults with an average total heat production of 200 W/person in volume and in mass?

1. In volume, the total heat production of 30 people is

$$THP = 200\left(\frac{\text{J}}{\text{s}\cdot\text{person}}\right)\times 30\ (\text{person})\times\left(\frac{1\ \text{kJ}}{1\,000\ \text{J}}\right)$$

$$= 6\left(\frac{\text{kJ}}{\text{s}}\right)$$

From Equation 7-6:

$$Q_{CO_2} = \frac{THP}{24\ 600} = \frac{6\ (\text{kJ/s})}{24\ 600\left(\frac{\text{kJ}}{\text{m}^3\ \text{of CO}_2}\right)} = 0.000\ 244\left(\frac{\text{m}^3\ \text{of CO}_2}{\text{s}}\right)$$

or

$$Q_{CO_2} = 0.878\left(\frac{\text{m}^3\ \text{of CO}_2}{\text{h}}\right)$$

2. In mass, the CO_2 production rate is

$$m_{CO_2} = Q_{CO_2}\rho_{CO_2} = 1.61\left(\frac{\text{kg of CO}_2}{\text{h}}\right)$$

Module 8:
Mass and Energy Balance for Ventilated Airspaces

In this module, we will discuss the following topics:

- Basic equations for mass and energy conservations.
- Sensible heat balance or temperature balance (energy conservation).
- Moisture balance (mass conservation-for water vapor).
- Contaminant balance (mass conservation-for specific gases such as ammonia, carbon dioxide, or dust).

Consider a ventilated building airspace. Air from outside coming into a room has a mass flow rate, m_o, enthalpy, h_o (including sensible heat and latent heat, h_{so} and h_{wo}), and contaminant concentration, C_o. Air being exhausted from the building has a mass flow rate, m, enthalpy, h (including sensible heat and latent heat, h_s and h_w), and contaminant concentration, C. When outside air enters the room, it is assumed completely mixed with the room air. Therefore, the room air has the same properties as the exhaust air. Furthermore, within the room airspace, the net heat transfer rate is q_n, moisture production rate is m_w, and contaminant production rate is m_c.

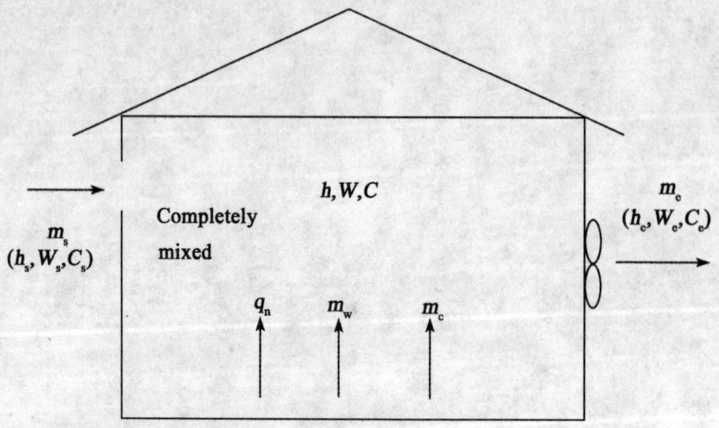

Figure 8-1. Heat and mass transfer of a ventilated airspace

Under steady-state and complete mixing conditions, the mass and energy conservation for the system described in Fig 8-1 are as following:

$$m_s h_s + q_n = m_e h_e \tag{8-1}$$

$$m_s W_s + m_w = m_e W_e \tag{8-2}$$

$$m_s = m_e = m \tag{8-3}$$

Where

m = the mass flow rate or production rate, kg/h

h = enthalpy of air, kJ/kg

Subscripts s and e refer to supply and exhaust, respectively.

8.1 Sensible Heat Balance Ventilation Requirement

Applying mass conservation to the supply and exhausted air, the rate of mass of air that enters the room (m_s or Q_s/v_s) must be equal to the mass of air that is exhausted out of the room (m_e or Q_e/v_e) under steady-state conditions.

$$m = \frac{Q_s}{v_s} = \frac{Q_e}{v_e} \tag{8-4}$$

Applying energy conservation to sensible heat, the total sensible heat contained in the supply air, plus the net sensible heat transfer rate of the building system, must be equal to the total sensible heat removed by the exhaust air under steady-state conditions. Substituting Equation 8-4 into Equation 8-1, gives

$$h_{ss}\frac{Q_s}{v_s} + q_n = h_{es}\frac{Q_e}{v_e} \tag{8-5}$$

Where h_{ss} and h_{es} are sensible heat for supply and exhaust air in kJ/kg of dry air, respectively. Solving Equation 8-5 gives volumetric ventilation rates *for sensible heat balance (or temperature balance) control* for exhaust air, Q_e, and supply air, Q_s in m^3/s, respectively:

$$Q_e = \frac{q_n v_e}{(h_{es} - h_{ss})} \tag{8-6}$$

$$Q_s = \frac{q_n v_s}{h_{es} - h_{ss}} \tag{8-7}$$

8.2 Moisture Balance Ventilation Requirement

Applying mass balance conservation to moisture, the total moisture contained in the supply air, plus the moisture produced in the room, must be equal to the total moisture removed by the exhaust air under steady-state conditions. Equation 8-2 can be rewritten as

$$W_s \frac{Q_s}{v_s} + m_w = W_e \frac{Q_e}{v_e} \tag{8-8}$$

Where W is the water content of air in kg of water vapor per kg of dry air (kg/kg). Subscripts "s" and "e" represent supply and exhaust air, respectively. And m_w is the water vapor production rate in kg/s within the airspace.

Solving Equation 8-8 gives volumetric ventilation rates *for moisture (or relative humidity) balance control* for exhaust air, Q_e, and supply air, Q_s in m^3/s, respectively:

$$Q_e = \frac{m_w v_e}{W_e - W_s} \tag{8-9}$$

$$Q_s = \frac{m_w v_s}{W_e - W_s} \tag{8-10}$$

8.3 Pollutant Balance Ventilation Requirement

When mass balance is applied to pollutant balance ventilation control, care should be taken on the units of pollutant concentration. Pollutant concentrations are commonly expressed in milligram of pollutant per cubic meter of air (mg/m^3) for solid and liquid pollutants, and in $\times 10^{-6}$ m^3/m^3 in volume or in mg/m^3 in mass for gaseous pollutants. From Equation 8-2, particulate pollutant mass balance at steady-state can be written as

$$C_s \frac{Q_s}{v_s} + m_c = C_e \frac{Q_e}{v_e} \tag{8-11}$$

$$Q_e = \frac{m_c v_e}{C_e v_e - C_s v_s} \tag{8-12}$$

$$Q_s = \frac{m_c v_s}{C_e v_e - C_s v_s} \tag{8-13}$$

Where C_e and C_s are the pollutant concentration of exhaust and supply air in kg of pollutant per m^3 of air, respectively. And m_c is the mass production rate of the particulate pollutant in kg/s.

For gaseous pollutants, concentrations are commonly expressed in part per million in volume (ppmv). Equation 8-12 and Equation 8-13 can be written as:

$$Q_e = \frac{m_c v_c v_e}{C_e v_e - C_s v_s} = \frac{Q_c v_e}{C_e v_e - C_s v_s} \tag{8-14}$$

$$Q_s = \frac{m_c v_c v_s}{C_e v_e - C_s v_s} = \frac{Q_c v_s}{C_e v_e - C_s v_s} \tag{8-15}$$

Where the gas concentrations C_e and C_s are in 10^{-6} m^3/m^3 air; m_c is the mass production rate for the gas in kg/s, and Q_c is the volumetric gas production rate in m^3/s; v_c is the specific volume of the gaseous pollutant in m^3/kg.

8.4 Minimum Ventilation Rates

The analysis above defines the minimum ventilation rates for the control of temperature, moisture and a specific pollutant. Minimum ventilation is a ventilation rate required to maintain the desirable room conditions (temperature, relative humidity and air quality) at given weather conditions. The minimum ventilation rate must be the highest of the three ventilation rates for balances of sensible heat, moisture and the given pollutant of concern at the design outside temperature. Therefore, to determine the minimum ventilation rate, one must calculate all ventilation requirements for balances of temperature, moisture and particular contaminant levels.

Because the sensible heat and moisture content are functions of air temperature, the ventilation requirements can be plotted versus the outside temperature at given indoor conditions, including the required temperature, relative humidity and an acceptable concentration of a pollutant such as carbon dioxide or particles. Such a plot is called a ventilation graph (Figure 9-1), in which the minimum ventilation rates (Q_{min}) are required to maintain acceptable indoor environmental conditions (temperature, humidity and a specific pollutant concentration) with respect to supply air temperatures. The minimum ventilation for an indoor environment should follow the uppermost portion of the three curves in a ventilation graph. Therefore, to determine the minimum ventilation rate, one must calculate all ventilation requirements for balances of temperature, moisture and specific pollutant concentration as described above, and select the highest of the three as the minimum ventilation. If the minimum ventilation rate is selected based on other humidity balance or pollutant concentration balance, supplemental heat is needed to maintain the desirable room temperature.

In practice, a ventilated airspace has a maximum ventilation capacity (Q_{max}). At the maximum ventilation, if the indoor temperature exceeds the required (usually the set-point) temperature, air-conditioning must be used to maintain the desirable temperature. If the relative humidity is too high at the maximum ventilation, dehumidification should be applied. If an air pollutant of concern exceeds the acceptable concentration, air cleaning equipment should be added to the ventilation system.

Example 8-1

There are 30 students in a classroom, each student produces 120 W of sensible heat and 100 g of water vapor per hour (or 70 W latent heat). The room air temperature is 20℃ and the specific volume is 0.84 m^3/kg. The carbon dioxide concentration of the supply air is 0.4×10^{-3} m^3/m^3 air. Supply air has a temperature of 0℃, specific volume of 0.775 m^3/kg and relative humidity of 80%. Heat loss through the building shelter is 2 kW and heat gain from equipment and lights are 1.5 kW. If we want to maintain an indoor environment in the classroom with a carbon dioxide concentration of less than 0.8×10^{-3} m^3/m^3 air, relative humidity of less than 50% and room temperature of 20℃, what should the minimum ventilation be for the supply air? At the chosen minimum ventilation rate, what are the actual temperature, relative humidity and carbon dioxide in the room, if there are no heating and humidification? Atmospheric conditions are assumed to be at sea level and standard, and the room air is completely mixed.

Solution

From the Psychrometric Chart 1, we can find the psychrometric properties of the supply air (0℃ and 80% RH) and exhaust air (20℃ and 50% RH) as follows:

$W_s = 0.003$, $v_s = 0.773\ m^3/kg$, $h_{ss} = 0\ kJ/kg$

$W_e = 0.0073$, $v_s = 0.84\ m^3/kg$, $h_{es} = 20\ kJ/kg$

The specific volume of CO_2 at 20℃, $v_c = 0.546\ m^3/kg$.

In order to determine the ventilation requirement, all three ventilation rates for sensible heat balance, moisture balance and CO_2 balance need to be calculated.

Ventilation for sensible heat balance, using the supply air rate, from Equation 7-3,

$$q_n = q_{oc} + q_b + q_e + q_r = 0.12 \times 30 - 2 + 1.5 = 3.1\ (kW)$$

$$Q_s = \frac{q_n v_s}{(h_{es} - h_{ss})} = \frac{3.1 \times 0.773}{20 - 0} = 0.12\left(\frac{m^3}{s}\right)$$

Ventilation for moisture balance, using the supply air rate, from Equation 8-13,

$$m_w = 0.1\left(\frac{kg}{person \cdot h}\right) \times 30\ person \times \frac{1\ h}{3\ 600\ s} = 0.000\ 83\ \frac{kg}{s}$$

$$Q_s = \frac{m_w v_s}{(W_e - W_s)} = \frac{0.000\ 83 \times 0.773}{0.007\ 3 - 0.003} = 0.149\ \left(\frac{m^3}{s}\right)$$

Ventilation for moisture balance, using the supply air rate, from Equation 8-13 and Equation 8-14,

$$Q_{cp-CO_2} = m_c v_c = \frac{THP}{24\ 600} = \frac{(0.12 + 0.07) \times 30}{24\ 600} = 0.000\ 232\ \left(\frac{m^3\ of\ CO_2}{s}\right)$$

$$Q_s = \frac{Q_c v_s}{C_e v_e - C_s v_s} \times 10^6 = \frac{0.000\ 232 \times 0.773}{0.8 \times 10^{-3} \times 0.84 - 0.4 \times 10^{-3} \times 0.773} = 0.494\ \left(\frac{m^3}{s}\right)$$

Since the highest ventilation requirement is for CO_2 balance control, the minimum supply air flow rate should be 0.494 m^3/s. At this ventilation rate, the actual room temperature and relative humidity, assuming there are no heating nor humidification, can be determined by solving the new sensible heat and moisture content in the room when $Q_s = 0.494\ m^3/s$. From Equation 8-5 and Equation 8-8,

$$h_{es} = \frac{q_n v_s}{Q_s} + h_{ss} = \frac{3.1 \times 0.773}{0.494} + 0 = 4.6 \left(\frac{\text{kJ}}{\text{kg}}\right)$$

$$W_e = \frac{m_w v_s}{Q_s} + W_s = \frac{0.000\,83 \times 0.773}{0.494} + 0.003 = 0.003\,2 \left(\frac{\text{kg of water}}{\text{kg of air}}\right)$$

From the psychrometric chart, we can read that the temperature is 4.6℃ and relative humidity is 61%.

8.5 Heating Capacity Calculation

8.5.1 Definition

Heating capacity is the supplemental heat required to maintain the room set-point temperature with the minimum ventilation rate at the design temperature (coldest outside temperature).

8.5.2 Applying Energy Balance

Assume minimum ventilation rate, m, has been determined, then the new sensible heat balance can be rewritten by taking the supplemental heat, h_s, into count:

$$m \cdot h_{ss} + q_n + q_s = m \cdot h_s$$

$$q_s = m(h_s - h_{ss}) - q_n$$

Where

m = minimum mass ventilation flow rate

q_s = heating capacity (supplemental heat required)

h_{ss} = sensible heat of supply air

h_s = sensible heat of room air

Note: If q_s is negative, then the supplemental heat becomes negative, indicating cooling is needed, and q_s becomes the cooling capacity.

8.6 Case Study: Ventilation Design for a Vehicle Cabin

8.6.1 Problem Description

You are working in the Research and Development (R/D) Department for the CDC Corporation. One day, CDC receives an order for an off-road service vehicle from the Department of Transportation of Alaska. The R/D chief engineer realizes that with the unique cold climate and harsh conditions, the existing CDC products need to be redesigned. He assigns you to design the ventilation system of the vehicle compartment. Following are some specifications of the vehicle:

- Battery operated (You must minimize the energy use!)
- Vehicle compartment is similar to a passenger car, 4.2 m^3 (2 m × 1.5 m × 1.4 m) in volume
- Total compartment surface area of 15 m^2
- Average insulation value of the compartment $RSI = 1.2\ m^2 \cdot ℃/W$
- Maximum capacity: 5 people

Inside conditions:

- Temperature, $T = 20℃$, (assume at this temperature, approximately 50% of the total heat produced by people is sensible heat)
- Relative humidity (Φ) less than 70%
- Carbon dioxide concentration, C, is less than $3 \times 10^{-3}\ m^3/m^3$ air in volume

Working conditions:

- Outside temperature (T_o) is $-40℃$
- Relative humidity (Φ_o) is 90%
- Carbon dioxide concentration (C_o) is $0.35 \times 10^{-3}\ m^3/m^3$ air in Volume

8.6.2 Find

- The air flow rate for the first setting of the fan on the front panel (i. e., minimum ventilation rate) to satisfy the inside conditions: temperature, relative humidity and carbon dioxide concentration
- Design heater size
- Construct a ventilation graph

8.6.3 Discussion

If C_{CO_2} inside is 2×10^{-3} m^3/m^3 air, what should Q_{min} be?

If only one person is inside, what will the T_i be if there is no heater?

8.6.4 Case Study: Solution

To meet the inside design conditions, you need to look at all ventilation requirements.

Step 1: Sensible heat balance (or temperature control) ventilation

$$Q_s = \frac{q_n v}{h_s - h_{so}}$$

From a Psychrometric Chart:

$$h_s = 20\ \frac{\text{kJ}}{\text{kg}} \qquad W = 10.2\ \frac{\text{g}}{\text{kg}}$$

$$h_{so} = -40\ \frac{\text{kJ}}{\text{kg}} \qquad W_o = 0.1\ \frac{\text{g}}{\text{kg}}$$

$$v = 0.842\ \frac{\text{m}^3}{\text{kg}} \qquad v_s = 0.66\ \frac{\text{m}^3}{\text{kg}}$$

$$q_n = q_{as} + q_b + q_l + q_e + q_m + q_r$$

$$q_{as} = 40\ \frac{\text{W}}{\text{m}^2} \times 1.8\ \frac{\text{m}^2}{\text{person}} \times 5\ \text{person} = 360\ \text{W}$$

$$q_b = \frac{A}{RSI}(t_o - t_i) = \frac{15}{1.2} \times (-60) = -750\ \text{W} \qquad (\text{“}-\text{” here is heat loss})$$

$$q_n = q_a + q_b = -390\ \text{W} = \left(-0.39\ \frac{\text{kJ}}{\text{S}}\right)$$

$$Q_s = \frac{-0.39\ \frac{\text{kJ}}{\text{s}} \times 0.842\ \frac{\text{m}^3}{\text{kg}}}{20\ \frac{\text{kJ}}{\text{kg}} - \left(-40\ \frac{\text{kJ}}{\text{kg}}\right)} = \frac{-0.328\ \frac{\text{kJ}\cdot\text{m}^3}{\text{s}\cdot\text{kg}}}{60\ \frac{\text{kJ}}{\text{kg}}} = -0.005\ 47\ \frac{\text{m}^3}{\text{s}} = -5.47\ \frac{\text{L}}{\text{s}}$$

The“ – ” sign in the ventilation rate indicates that even at zero ventilation, the desired temperature (20℃) could not be maintained. Supplemental heat must be added.

Step 2: Moisture balance ventilation

From the relationship of latent heat and moisture content in the air,

$$h_w = W(2\ 501 + 1.805T)$$

Where the term $(2\ 501 + 1.805T)$, in kJ/kg of water, is the specific heat of water vapor at the temperature T. The latent heat production rate, q_w in kJ/s and the moisture production rate, m_w in kg of water vapor per s, can be written as

$$q_w = m_w(2\ 501 + 1.805T)$$

Because the latent heat is 50% of the total heat, q_w = 0.36 kJ/s, the moisture production rate is

$$m_w = \frac{q_w}{2\ 501 + 1.805T} = \frac{0.36\ \frac{\text{kJ}}{\text{s}}}{(2\ 501 + 1.805 \times 20)\ \frac{\text{kJ}}{\text{kg}}} = 0.000\ 142\ \frac{\text{kg}}{\text{s}}$$

Then from Equation 8-10, gives the ventilation rate of the supply air for moisture balance

$$Q_s = \frac{m_w v_s}{(W_e - W_s)}$$

$$= \frac{0.000\ 142\ \frac{\text{kg}}{\text{s}} \times 0.66\ \frac{\text{m}^3}{\text{kg}_{\text{air}}}}{0.001\ 02\ \frac{\text{kg}}{\text{kg}_{\text{air}}} - 0.000\ 1\ \frac{\text{kg}}{\text{kg}_{\text{air}}}} = 0.011\ 8\ \frac{\text{m}^3}{\text{s}}$$

Step 3: CO_2 balance ventilation (exhaust)

The volumetric CO_2 production rate can be calculated from the total heat production,

$$Q_{co_2} = \frac{THP(\frac{\text{kJ}}{\text{s}})}{24.6\ \frac{\text{kJ}}{\text{L}}} = \frac{0.72\ \frac{\text{kJ}}{\text{s}}}{24.6\ \frac{\text{kJ}}{\text{L}}} = 0.029\ 3\ \frac{\text{L}}{\text{s}} = 0.000\ 029\ 3\ \frac{\text{m}^3}{\text{s}}$$

From Equation 8-15, the ventilation rate of the supply air for the CO_2 balance is

$$Q_s = \frac{Q_{co_2} v_s}{C_e\ v_e - C_s\ v_s} \times 10^6$$

$$= \frac{0.000\,029\,3\ \frac{m^3_{co_2}}{s} \times 0.66\ \frac{m^3}{kg}}{3 \times 10^{-3} \frac{m^3}{m^3_{air}} \times 0.842\ \frac{m^3}{kg} - 0.35 \times 10^{-3} \frac{m^3}{m^3_{air}} \times 0.66\ \frac{m^3}{kg}}$$

$$= 0.011\,1\ \frac{m^3}{s}$$

The minimum ventilation Q_{min}, should be the highest of three ventilation balances, i. e. , moisture balance ventilation rate, 0. 011 8 m^3/s.

8.7 Problems

8.7.1 Given

- A central Illinois animal building is maintained at 18℃, less than or equal to 70% relative humidity, and less than or equal to 3×10^{-3} m^3/m^3 air in volume carbon dioxide.
- On a cold day, outside temperature dropped to −20℃ and 80% relative humidity.
- Deep floor temperature is approximately 10℃ year around.
- The building houses 1 200 pigs averaged at 25 kg each.
- Carbon dioxide concentration of outside air is 0.35×10^{-3} m^3/m^3 air in volume. The density of carbon dioxide is 1. 53 times of the air at the given temperature.
- Carbon dioxide production from animals can be calculated based on their total heat production for it is basically a metabolic product. On average, 1 liter CO_2 for each 24. 6 kJ of total heat produced.
- Building elevation is approximately at sea level.
- Building specifications are as following:

Dimension: 50 m long, 12 m wide and 3 m high (floor to ceiling), rectangular.

Insulation:

- walls with R20 insulation, $RSI = 3.94\ m^2 \cdot ℃/W$
- ceiling with R30 insulation, $RSI = 5.48\ m^2 \cdot ℃/W$
- concrete floor, $RSI = 1.76\ m^2 \cdot ℃/W$

- concrete foundation with 50 mm polyethylene rigid insulation and 300 mm below exterior grade, $F=0.4\ W/(m\cdot ℃)$

- Heat produced by other equipment, such as motors, is 500 W.
- Light: 30 lamps, 150 W each.
- Solar radiation is negligible.

8.7.2 Find

1. Heating load for the building. Heating load is the net heat loss or gain of the building airspace. It is the sum of the heat transfer through the building shelter, sensible heat produced by animals, equipment and lights, etc.

2. Calculate the production rate of carbon dioxide by animals within the building.

3. On that cold day (−20℃ and 80% of RH), you enter the room and measure that the actual relative humidity of the room is 60%, temperature 18℃ and CO_2 concentration $2\times10^{-3}\ m^3/m^3$ air in volume. A bank of exhaust fans exhausts 3 000 liters per second of room air. At this ventilation rate, calculate:

a. sensible heat removed through ventilation

b. moisture removed through ventilation

c. carbon dioxide removed through ventilation

8.8 Laboratory Exercise: Measuring Ventilation Rate in an Airspace

8.8.1 Objectives

- To determine the ventilation rate of the airplane cabin mock-up based on airborne pollutant balance and air velocity measurements.
- To apply mass and energy conservation laws in a real world system.
- Learn how to conduct tracer gas study and operate a hot-wire anemometer.

Summary:

Ventilation rate in a room will be measured using two common methods: (1) tracer gas method and (2) hot-wire anemometry. In the tracer gas method, a known amount of gas (CO_2) is fed into the room and the concentration at the outlet is measured. In this lab you will use the tracer gas method to measure the ventilation rate of the Boeing aircraft cabin in the BESS Lab. Heating pads and a carbon dioxide tank will be used to introduce heat and a pollutant into the cabin. Measurements of CO_2, temperature and velocity will be taken from the air exhaust. Then apply the energy and mass conservation to determine the ventilation rate of the cabin and heat transfer through the structure. Students will take velocity measurements at the air outlets using a hot-wire anemometer and do a comparison.

8.8.2 Equipment

1. Boeing airplane cabin mock-up
2. 28 heating pads attached to mannequins, each is 53 W (total: $28 \times 53 = 1\,484$ W)
3. Carbon dioxide tank, adjusted to a flow rate of 4 L/min
4. 1 portable fan
5. Hand held anemometer/thermometer
6. CO_2 meter and desktop computer
7. Stopwatch and tape measure

8.8.3 Procedure

A. CO_2 tracer gas method:

1. Use the thermometer and CO_2 meter to measure ambient air temperature and C_{CO_2} concentration (let the meter warm up for at least 3 minutes before reading).
2. Turn on the cabin ventilation (TA assist).
3. Put the mixing fan in the cabin, plug in and turn on, directing air to ensure best mixing.
4. Close the cabin door.
5. *At the same time*:

a. Plug in heating pads (2 plugs near sink, 2 near garage door).

b. Open CO_2 valve and adjust regulator to 80 kPa (TA assist).

c. Start the stopwatch.

6. Run the system for 5 minutes, then measure temperature and C_{CO_2} at exhaust outlet.
7. Repeat CO_2 measurement at intervals of 2 minutes until stabilized.

B. Velocity measurement method:

You will use the hot wire anemometer to measure velocity at the outlets and inlet for comparison.

Outlets:

1. Measure the cross sectional area of the air outlets.
2. Using the hot-wire anemometer, measure the air velocity at six points over the outlet area. Make sure the probe is at least 2 inches away from the outlet wall.
3. Repeat the measurement for three times on each outlet.

Inlet:

1. Measure the cross sectional area of the air inlet.
2. Measure the air velocity at 9 points in a 3 by 3 grid across the face of the inlet.

Lab Report Instructions: Follow the lab report format and be sure to complete and include the following elements. Include in your Methods/Approach section:

1. A derivation of the ventilation rate through the cabin based on conservation of mass (CO_2).

Conservation of Mass at steady-state:

$$0 = \sum m_{\text{in}} - \sum m_{\text{out}} \qquad \text{units} = \mu g/s$$

2. A derivation of the heat transfer rate through the cabin walls based on conservation of energy.

Conservation of Energy at steady state:

$$0 = \sum q_{\text{in}} - \sum q_{\text{out}} \qquad \text{units} = J/s = W$$

Hint: Use the following system diagram of the aircraft cabin as your starting point.

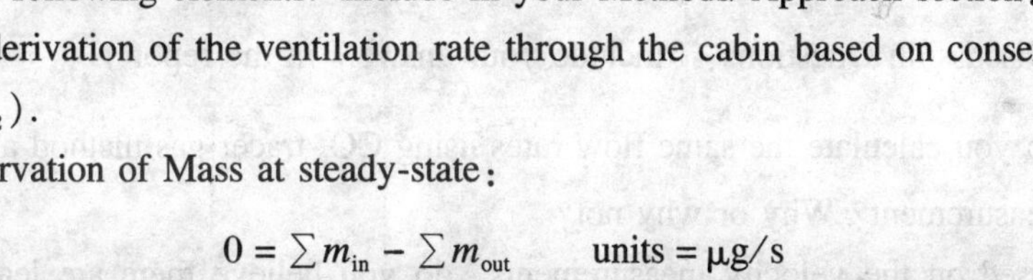

Where:

$m_{CO_2,\ in}$ = mass flow rate of CO_2 into the room from the supply tank (μg/s)

C_{in} and C_{out} = Concentration of CO_2 in the inlet and outlet, respectively ($\mu g/m^3$)

Q_{in} and Q_{out} = Air flow in the inlet and outlets, respectively ($Q_{in} = Q_{out}$) (m^3/s)

q_{gen} = heat generated in the room

8.8.4 Find

1. The ventilation flow rate using the tracer gas method of CO_2 measurement.
2. The ventilation flow rate using the velocity measurement method based on:
   a. air velocity at the outlets
   b. air velocity at the inlet
3. Using the calculated ventilation from 1, calculate the heat transfer rate through the cabin shelter.

8.8.5 Discussion Questions: (Include your answers in the report)

1. Do you calculate the same flow rates using CO_2 tracer gas method and velocity measurement? Why or why not?
2. Based on the velocity measurements, do you believe there are leaks through the walls of the air craft? Explain your answer.
3. What are your sources of error in the tracer gas measurement? How did the background levels of CO_2 impact the experiment?
4. What sources of error might come into play when calculating the heat transfer through the shelter? Where all heat sources accounted for in the system?

8.8.6 Useful conversion

$$C_{i,\mu g/m^3} = \frac{PM_i}{RT} C_{i,\ mg/L}$$

Where: C_i = concentration of species i

P = atmospheric pressure, Pa

T = absolute temperature, K

M_i = molar mass of species i

R = gas constant: 8.314 J/(K · mol)

Module 9: Ventilation Graph

A ventilation graph is a graph in which the minimum ventilation rates required to maintain an acceptable indoor environment are plotted against outside temperatures. A ventilation graph is determined by the following variables:

- Building shelter insulation levels.
- Production of heat, moisture and contaminants within the building airspace.
- Room conditions—set-points of temperature, relative humidity and pollutant concentration.
- Outside conditions—temperature, relative humidity and pollutant concentration.

Take a ventilation rate of exhaust air, Q_e, for example. If a room is designed to have a set-point temperature, T, a relative humidity, Φ, and a pollutant concentration, C_c, it is said the room should be maintained at the given temperature T, relative humidity is no higher than Φ, and the specific pollutant concentration is no higher than C_c. Using Equation 8-6, Equation 8-9 and Equation 8-12, plotting the ventilation versus the outside temperature, there will be three curves as sketched in Figure 9-1.

A ventilation system should always provide air delivery at or above the upper-most curve. As the example in Figure 9-2 shows, the ventilation rate should follow the contaminant and moisture balance requirements when outside temperatures are below -21℃ and 9℃, respectively. During cold weather, under-ventilation (shaded area in Figure 9-2) will cause either high humidity or air quality problems, or both. Over-ventilation will increase supplemental heat consumption. In an extreme situation, over-ventilation in cold weather may lead to temperature control failure if heater capacity is inadequate.

The minimum ventilation for an indoor environment should follow the uppermost portion of the three curves in a ventilation graph. Therefore, to determine the minimum ventilation rate, all ventilation requirements must be calculated for balances of temperature, moisture and specific pollutant concentration as described above, and

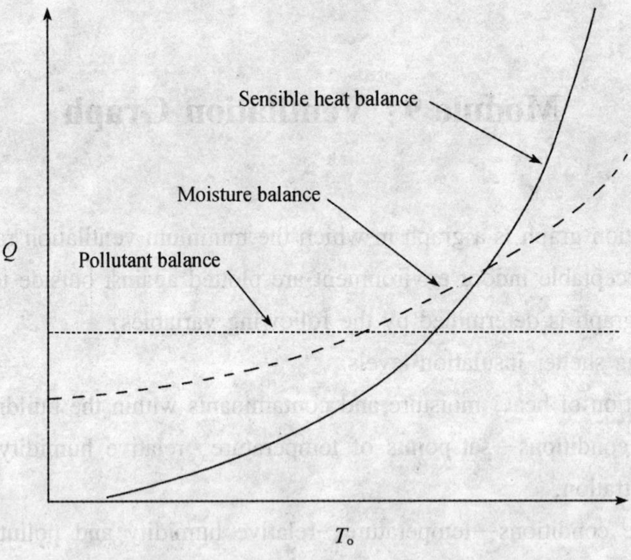

Figure 9-1. A ventilation graph - ventilation rates required to maintain the balance of room temperature, humidity and pollutant concentration at different outside temperatures

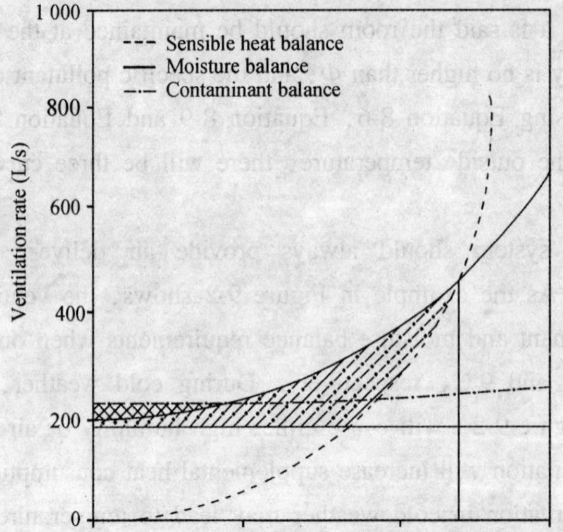

Figure 9-2. Example of a ventilation graph and heat deficit temperature, T_d. Shaded areas show insufficient ventilation for moisture and/or contaminant balance control. The minimum ventilation should follow the uppermost curves

then select the highest of the three as the minimum ventilation. If the minimum ventilation rate is selected based on humidity balance or pollutant concentration balance, supplemental heat is needed to maintain the desirable room temperature.

9.1 Heat Deficit Temperature

Heat deficit temperature (T_d) is an outside temperature below which supplemental heat is required to maintain acceptable relative humidity levels or air quality in the building. In Figure 9-2, the heat deficit temperature is 9℃. In any building, the heat deficit temperature changes with the total heat and moisture production, set-point temperature, and relative humidity. For example, if a room is partially filled with animals, the heat deficit temperature will be higher than if the room is full of animals because the animal heat production is lower; supplemental heat is needed even though the outside temperature may be fairly high. Table 9-1 gives heat deficit temperatures and heating seasons for the prairies based on rooms filled to capacity with animals.

Table 9-1. Heat deficit temperatures for swine buildings at given room temperatures, with typical insulation levels and a full housing density of animals

Swine Buildings	Room Temperature (℃)	Heat Deficit (℃)
Dry Sow	18	−10
Farrowing	18*	5
Weanling		
7 kg	28	15
20 kg	24	10
Grower/Finisher		
25 kg	23	8
50 kg	16	−20
90 kg	15	−26

* *Room temperature only. Heat lamps or other supplemental heat should be provided to piglets.*

Heating seasons vary largely with geographical location and weather. To determine heating seasons, local weather data must be used. The heating seasons listed in Table 9-2 are based on the typical meteorological year (TMY) files for Saskatoon, Canada. The joint frequency distribution for temperature and relative humidity for Saskatoon is shown in Figure 9-3. It can be used as representative weather throughout the prairies. For each outside temperature, relative humidity is given two values: mean (average) and 95%. The 95% value of relative humidity indicates that at a given temperature, the given relative humidity will not be exceeded 95% of the time. For example, at −35℃, the relative humidity would not exceed 80% ninety-five percent of the time.

Table 9-2. Heat deficit temperatures and prairie heating seasons for buildings at given room temperatures

Swine Building	Room temperature (℃)	Head deficit temperature (℃)	Heating season (d/year)
Dry Sow	18	−10	90
Farrowing	18*	5	210
Weanling			
7 kg	28	15	280
20 kg	24	10	245
Grower/Finisher			
25 kg	23	8	230
50 kg	16	−20	30
90 kg	15	−26	12

** Room temperature only. Heat lamps or other supplemental heat should be provided to piglets.*

9.2 Maximum Ventilation Requirement

From the sensible heat balance ventilation, it is obvious that when the outside temperature approaches the inside temperature, the ventilation rate for sensible heat balance (Equation 8-4 and Equation 8-5) approaches infinity, which is impractical to achieve without air-conditioning. Therefore, a maximum ventilation rate must be set for fan selections.

Maximum ventilation is the actual ventilation rate that an air delivery system can provide for summer conditions. Without an air-conditioning system, maximum ventilation for an animal building is up to 60 times of its minimum ventilation rate. Another rule of thumb, animal buildings use 60 air changes per hour (ACH) as the maximum ventilation rate for selecting fans.

In practice, a ventilated airspace has a maximum ventilation capacity (Q_{max}). At the maximum ventilation, if the indoor temperature exceeds the required (usually the set-point) temperature, air-conditioning must be used to maintain the desirable temperature. If the relative humidity is too high at the maximum ventilation, dehumidification should be applied. If an air pollutant of concern exceeds the acceptable concentration, air cleaning equipment should be added to the ventilation system.

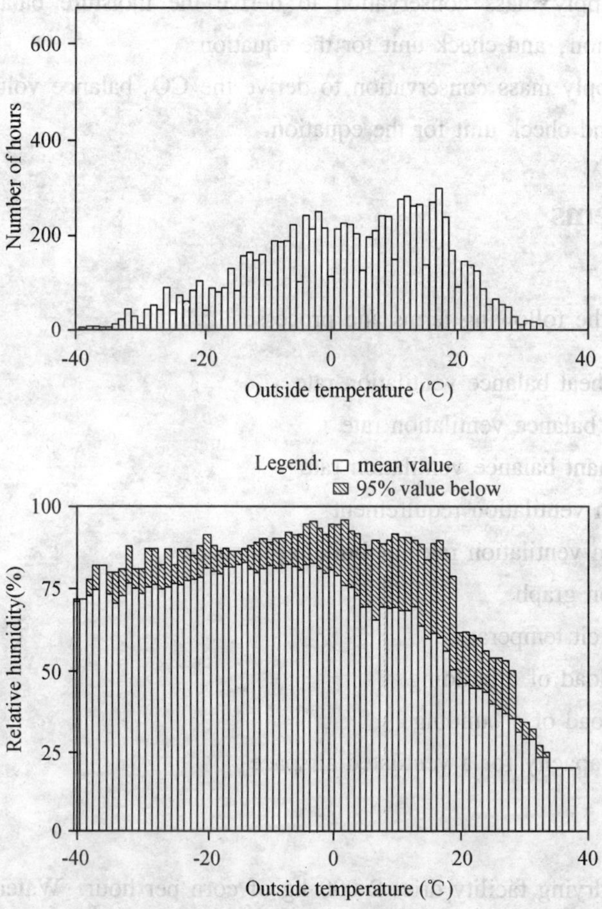

Figure 9-3. Joint frequency distribution for temperature and relative humidity in a cold and dry climate

9.3 Work Sheet: Ventilation Requirement

Objectives: To derive equations to calculate minimum ventilation rates based on moisture balance, carbon dioxide balance or any other specific contaminant concentration balances.

Task 1: Sketch the physical model of a ventilated airspace, list all variables involved.

Task 2: Apply mass conservation to derive the moisture balance volumetric ventilation equation, and check unit for the equation.

Task 3: Apply mass conservation to derive the CO_2 balance volumetric ventilation equation, and check unit for the equation.

9.4 Problems

9.4.1 Define the following terms and processes

a. Sensible heat balance ventilation rate

b. Moisture balance ventilation rate

c. Contaminant balance ventilation rate

d. Minimum ventilation requirement

e. Maximum ventilation rate for building airspace

f. Ventilation graph

g. Heat deficit temperature

h. Heating load of a building

i. Cooling load of a building

j. Heating capacity for a ventilated airspace

9.4.2 Given

- A grain drying facility dries 2 000 kg of corn per hour. Water content of the corn is 30% in mass initially. After one hour drying, water content of the corn is 20% (All mass and percentage are on wet base).
- Outside air at 20℃ and 40% relative humidity. Air is heated (sensible heating) and then is forced into the drying chamber by a blower.
- On average, air within the drying chamber is maintained at 45℃ and 35% relative humidity.
- Assume the facility is well insulated and the heat transfer through the chamber shelter is negligible.

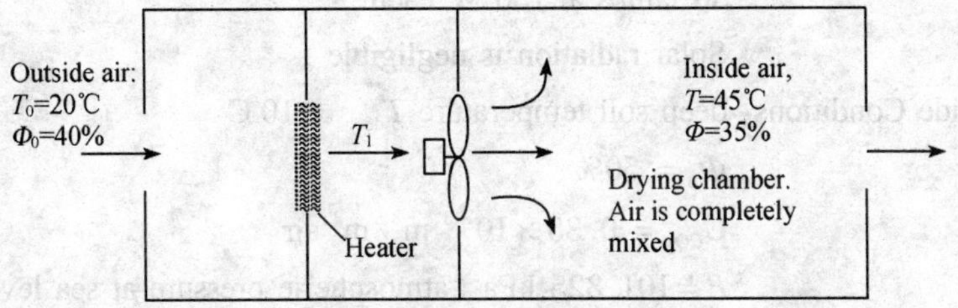

9.4.3 Find

a. Total moisture removed per hour from the chamber.

b. Air flow rate of the blower based on moisture balance ventilation.

c. Temperature of supply air after heater but before entering the drying chamber, T_1 (*Hint: The enthalpy after the heater is the same as the enthalpy in the drying chamber*).

d. The heater size (i. e., heating capacity), q_s, (in kW) of the drying facility.

Note: The heater only gives off sensible heat.

9.5 Case Study: Ventilation Requirement

9.5.1 Given

Animal Building:

Condition: T_i = 18℃, RH_i = 70%

$C_{iCO_2} = 3\times10^{-3}$ m³/m³ air

Note: $\rho_{CO_2} = 1.53\cdot\rho_{air}$ (at T_i)

Animals: 1 200 pigs at 25 kg each

CO_2 production of animals = 1 L CO_2/ 24.6 kJ total heat produced

Dimensions: 50 m long, 12 m wide and 3 m high to ceiling

Walls: R20 insulation at RSI = 3.94 m²·℃/W

Ceiling: R30 insulation at RSI = 5.48 m²·℃/W

Floor: RSI = 1.76 m²·℃/W

Perimeter: F = 0.4 W/(m·℃)

Heat Generated: 500 W by equipment

30 lamps at 150 W each

Solar radiation is negligible

Outside Conditions: deep soil temperature $T_{deep} = 10℃$

$\Phi_o = 50\%$

$C_{CO_2} = 0.35 \times 10^{-3}\ m^3/m^3$ air

$P = 101.325$ kPa (atmospheric pressure at sea level)

9.5.2 Find

(1) Generate a ventilation graph for outside temperature range of $-20℃$ to $10℃$. Assume outside relative humidity is constant at 50%. Use a spreadsheet or compute manually using at least 7 points.

(2) Maximum ventilation rate (i.e., total fan capacity).

(3) Minimum ventilation rate to maintain inside conditions when outside temperature is:

$$T_o = -20℃$$

$$T_o = 5℃$$

4. Heat deficit temperature.
5. Heater size for the building.

9.5.3 Solution

To get the minimum ventilation rate based on sensible heat balance, we use the Equation 8-6:

$$Q_e = \frac{q_n v_e}{(h_{es} - h_{ss})}$$

To compute the net heat transfer of the building, q_n, we use the following equation:

$$q_n = q_{as} + q_b + q_l + q_e + q_m + q_r$$

The individual heat transfer rate will be computed as follows:

(1) The sensible heat produced by animals can be computed using the equation:

$$q_{as} = (SHP) \times (\text{weight of pigs}) \times (\text{No. of pigs})$$

From Table 7-4, for 25 kg pigs at 18℃, the *SHP* can be calculated by interpolation:
At 18℃,

$$SHP_{20\text{ kg}} = \frac{(2.3-3.0)}{(20-15)}(18-15)+3.0=2.58(\text{W/kg})$$

$$SHP_{40\text{ kg}} = \frac{(1.6-2.0)}{(20-15)}(18-15)+2.0=1.76(\text{W/kg})$$

$$SHP_{25\text{ kg}} = \frac{(1.76-2.58)}{(40-20)}(25-20)+2.58=2.375(\text{W/kg})$$

$$\text{Therefore}, q_{as} = \left(2.375\ \frac{\text{W}}{\text{kg}}\right)\left(25\ \frac{\text{kg}}{\text{pig}}\right)(1\ 200\ \text{pigs}) = 71\ 250(\text{W})$$

(2) The heat transfer through the building shelter is:

$$q_b = q_{wall} + q_{ceiling} + q_{floor} + q_{perimeter}$$

Using the equations for heat transfer through walls, ceiling, floor, and perimeter:

$$q_{wall} = \frac{A_{wall}(\text{m}^2)}{RSI_{wall}\left[\frac{\text{m}^2\cdot℃}{\text{W}}\right]}(T_i - T_o)(℃)$$

$$q_{ceiling} = \frac{A_{ceiling}(\text{m}^2)}{RSI_{ceiling}\left[\frac{\text{m}^2\cdot℃}{\text{W}}\right]}(T_i - T_{attic})(℃)$$

$$q_{floor} = \frac{A_{floor}(\text{m}^2)}{RSI_{floor}\left[\frac{\text{m}^2\cdot℃}{\text{W}}\right]}(T_i - T_{ground})(℃)$$

$$q_{perimeter} = P_{building}(\text{m}) \times F\left[\frac{\text{W}}{\text{m}\cdot℃}\right] \times (T_i - T_o)(℃)$$

(3) To get v, h_s and h_{so}, we use the psychrometric chart since atmospheric pressure is at sea level:
To get the minimum ventilation rate based on moisture balance, we use the equation:

$$Q_w = \frac{m_w v}{(W - W_o)}$$

From Table 7-4, for 25 kg pigs at 18℃, moisture production rate, m_w can be calculated by interpolation:

$$\text{At } 18℃,\quad MP_{20\text{ kg}} = \frac{(1.03-0.86)}{(20-15)}(18-15)+0.86=0.962\ \frac{\text{mg}}{\text{kg}\cdot\text{s}}$$

$$MP_{40\ \text{kg}} = \frac{(0.61 - 0.53)}{(20 - 15)}(18 - 15) + 0.53 = 0.578\ \frac{\text{mg}}{\text{kg} \cdot \text{s}}$$

$$MP_{25\ \text{kg}} = \frac{(0.578 - 0.962)}{(40 - 20)}(25 - 20) + 0.962 = 0.866\ \frac{\text{mg}}{\text{kg} \cdot \text{s}}$$

We can then compute for m_w:

$$m_w = \left(0.866\ \frac{\text{mg}}{\text{kg} \cdot \text{s}}\right)\left(25\ \frac{\text{kg}}{\text{pig}}\right)(1\ 200\ \text{pigs}) = 25\ 980\ \frac{\text{mg}}{\text{s}} = 25.98\ \text{g}\ \frac{H_2O}{\text{s}}$$

To get v, W and W_o, we use the psychrometric chart since atmospheric pressure is at sea level:

To get the minimum ventilation rate based on carbon dioxide balance, we use Equation 8-12:

$$Q_e = \frac{m_c v_e}{(C_e v_e - C_s v_s)}$$

In order to compute for m_c, we have to get the total heat production of the animals first.

From Table 7-4, for 25 kg pigs at 18℃, the *THP* can be calculated by interpolation:

At 18℃,
$$THP_{20\ \text{kg}} = \frac{(4.8 - 5.0)}{(20 - 15)}(18 - 15) + 5.0 = 4.88\ \frac{\text{W}}{\text{kg}}$$

$$THP_{40\ \text{kg}} = \frac{(3.1 - 3.3)}{(20 - 15)}(18 - 15) + 3.3 = 3.18\ \frac{\text{W}}{\text{kg}}$$

$$THP_{25\ \text{kg}} = \frac{(3.18 - 4.88)}{(40 - 20)}(25 - 20) + 4.88 = 4.455\ \frac{\text{W}}{\text{kg}}$$

Therefore, $q_t = \left(4.455\ \frac{\text{W}}{\text{kg}}\right)\left(25\ \frac{\text{kg}}{\text{pig}}\right)(1\ 200\ \text{pigs}) = 133\ 650\ \text{W}$

$$q_t = 133.65\ \text{kW}$$

From Equation 7-6, CO_2 production rate is

$$Q_p = \left(\frac{1\ \text{L}\ CO_2}{24.6\ \text{kJ}}\right)\left(133.65\ \frac{\text{kJ}}{\text{s}}\right) = 5.43\ \frac{\text{L}\ CO_2}{\text{s}}\ \text{or}\ 19\ 548\ \frac{\text{L}}{\text{h}}$$

We also have, $C = \dfrac{0.003\ \text{L}\ CO_2}{\text{L air}}$

$$C_s = \frac{0.000\,35 \text{ L } CO_2}{\text{L air}}$$

Therefore,

$$Q_e = \frac{\left(5.43 \dfrac{\text{L } CO_2}{\text{s}}\right)}{(0.003 - 0.000\,35)\dfrac{\text{L } CO_2}{\text{L air}}} = 2\,049.1 \frac{\text{L air}}{\text{s}} = 2.05 \frac{\text{m}^3}{\text{s}}$$

The equations presented can then be used to set-up MS Excel.

Table 9-3. An example of spreadsheet data for a ventilation graph using MS Excel

	(W)							
	-20℃	-15℃	-10℃	-5℃	0℃	5℃	10℃	
q_w	3 587.82	3 115.74	2 643.65	2 171.57	1 699.49	1 227.41	755.33	
q_c	3 941.61	3 394.16	2 846.72	2 299.27	1 751.82	1 204.38	656.93	
q_f	2 727.27	2 727.27	2 727.27	2 727.27	2 727.27	2 727.27	2 727.27	
q_p	1 884.80	1 636.80	1 388.80	1 140.80	892.80	644.80	396.80	
q_b	12 141.50	10 873.97	9 606.44	8 338.92	7 071.39	5 803.86	4 536.34	
q_l	4 500	4 500	4 500	4 500	4 500	4 500	4 500	
q_m	500	500	500	500	500	500	500	
q_{as}	71 250	71 250	71 250	71 250	71 250	71 250	71 250	
q_n	64 108.50	65 376.03	66 643.56	67 911.08	69 178.61	70 446.14	71 713.66	(W)
v	0.84	0.84	0.84	0.84	0.84	0.84	0.84	(m^3/kg dry air)
h_s	18.11	18.11	18.11	18.11	18.11	18.11	18.11	(kJ/kg dry air)
h_{so}	-20.12	-15.09	-10.06	-5.03	0.00	5.03	10.06	(kJ/kg dry air)
Q_s	1.40	1.65	1.98	2.45	3.19	4.50	7.45	(m^3/s)
Q_s	5 047.11	5 926.73	7 120.51	8 833.31	11 497.68	16 211.55	26 817.77	(m^3/h)
m_w	25.98	25.98	25.98	25.98	25.98	25.98	25.98	(g H_2O/s)
W	9.00	9.00	9.00	9.00	9.00	9.00	9.00	(g H_2O/kg dry air)
W_o	0.30	0.50	0.80	1.25	1.90	2.70	3.80	(g H_2O/kg dry air)
Q_w	2.50	2.56	2.65	2.80	3.06	3.45	4.18	(m^3/s)
Q_p	5.43	5.43	5.43	5.43	5.43	5.43	5.43	(L CO_2/s)
C	0.003	0.003	0.003	0.003	0.003	0.003	0.003	(L CO_2/L air)
C_o	0.000 35	0.000 35	0.000 35	0.000 35	0.000 35	0.000 35	0.000 35	(L CO_2/L air)
Q_c	2.05	2.05	2.05	2.05	2.05	2.05	2.05	(m^3/s)

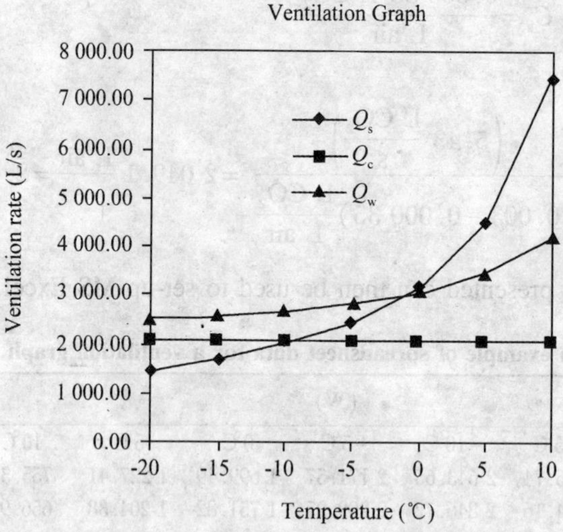

Figure 9-4. An example of a ventilation graph

The minimum ventilation rates can be determined from this table.

Table 9-4. Minimum ventilation rates determined from the ventilation graph

	-20℃	-15℃	-10℃	-5℃	0℃	5℃	10℃	
Q_s	1 401.98	1 646.31	1 977.92	2 453.70	3 193.80	4 503.21	7 449.38	(L/s)
Q_w	2 496.47	2 555.21	2 648.69	2 802.49	3 059.05	3 447.50	4 176.78	(L/s)
Q_c	2 049.06	2 049.06	2 049.06	2 049.06	2 049.06	2 049.06	2 049.06	(L/s)

1. The maximum ventilation rate is 30 times the minimum ventilation rate (at the design temperature of −20℃).

$$Q_{max} = 30 \times 2\,496.47\ \frac{L}{s} = 74\,894.1\ \frac{L}{s}$$

2. The minimum ventilation rate is the highest value of the three ventilation rates at a certain temperature:

$$T_o = -20℃: \qquad Q_{min} = 2\,496.47\ \frac{L}{s}$$

$$T_o = 5℃: \qquad Q_{min} = 4\,503.21\ \frac{L}{s}$$

3. The heat deficit temperature is the intersection point of the sensible heat balance ventilation line and the moisture balance ventilation line.

$$T_d = -2.0℃$$

4. The heater size for the building can be computed using the following equation,

$$q_s = m(h_s - h_{so}) - q_n$$

$$q_s = \left[\frac{2.5 \frac{\text{m}^3}{\text{s}}}{0.836 \frac{\text{m}^3}{\text{kg dry air}}} \times (18.1 + 20.1) \frac{\text{kJ}}{\text{kg dry air}} \right] - 64.11 \frac{\text{kJ}}{\text{s}} = 50 \text{ kW}$$

9.6 Laboratory Exercise: Fan Test Chamber

9.6.1 Objectives

To determine the performance in airflow and ventilation efficiency ratio of an agricultural propeller fan.

9.6.2 Summary

Fans are designed for many different applications. In order to determine the optimal fan for a particular job, the performance of each fan must be well understood. In this lab, we will use the fan test chamber in the BESS lab to determine the performance characteristics (airflow and airflow per watt of power ventilation efficiency ratio) of a fan at different static pressures and different voltages. A spreadsheet program will calculate airflow rate through the fan based on the measured pressure drop and psychrometric properties.

9.6.3 Equipment

1. Fan test chamber
2. Vista 1803P ventilation fan (or similar as specified by the Lab Manager)
3. Valhalla 2101 digital power analyzer
4. Strobe tachometer
5. Fan test spreadsheet and laptop
6. Psychrometer

9.6.4 Procedure

1. Lab Manager will explain the testing purpose and procedures

2. Make sure fan is properly secured and turn on the fan test chamber

3. Measure psychrometric room conditions and input into the spreadsheet

4. Adjust static pressure to 0 inches of H_2O to simulate "free air"

5. Adjust voltage to (230 ± 0.5) V (or as specified by the Lab Manager)

6. Measure the pressure drop across the nozzle bank, power usage, and angular fan speed (RPMs)

7. Input this data into the fan test spreadsheet

8. Repeat the process for static pressures of 0.05, 0.1, 0.15, 0.2, 0.25, and 0.3, including a final point with flow nozzles blocked for Airflow cutoff (no flow)

9.6.5 Find

1. airflow vs. static pressure curve

2. ventilation efficiency ratio vs. static pressure curve

Note: make sure you use SI units-m^3/h, W, and Pa

9.6.6 Questions to think about

1. Why do we test both airflow and ventilation efficiency ratio?

2. What applications would this fan be suitable for? Why? Think in terms of the fans pressure capabilities, construction materials, etc.

3. What are the relationships between RPM, static pressure and flow rate? Plotting these parameters two at a time may reveal any relationships. How does the static pressure affect the flow rate? RPM? What is the impact of this relationship?

4. What kinds of calculations do you think the spreadsheet is doing?

Module 10: Physiological Responses

In designing a heating, ventilation and air-conditioning system, it is important to consider the effect of the environmental conditions (temperature, humidity and pollutant concentrations) on the occupants. In this module, some typical physiological responses to certain environmental conditions will be introduced. Specifically,

- Human comfort zone
- Temperature set-points for animal facilities
- Heat stress for animals
- Airborne pollutant concentrations and human respiratory health
- Threshold limits for aerial pollutants

10.1 Human Thermal Comfort Zone

Thermal comfort of humans is determined by the heat transfer rate from skin to the environment. The heat transfer rate is a function of the ambient temperature, air velocity, relative humidity, clothing level (insulation level), and wetness of the skin. There are different descriptions for thermal comfort depending on outdoor or indoor conditions.

For outdoor conditions, wind chill factor (or wind chill index) is often used for cold weather and heat index for hot weather. Wind chill factor is an apparent temperature felt on the exposed skin, which is a function of the ambient air temperature and the wind speed. The wind chill temperature (often called the wind chill factor) is always lower than the ambient air temperature, except at higher temperatures where wind chill is considered less important. In cases where the apparent temperature is higher than the air temperature, the heat index is used instead. The heat index is an apparent temperature that combines ambient air temperature and relative humidity in an attempt to determine the human-perceived equivalent temperature. Heat index is always higher than the apparent temperature. The human body normally cools itself by perspiration for an evaporative cooling. However, when the relative humidity is high, the evaporation rate is reduced, so heat is removed from the body at a lower rate causing it to retain more heat than it would in dry air. In some literature, the heat index and the wind chill factor are denoted collectively by the single terms "apparent temperature" or "relative outdoor temperature" (ASHRAE, 2005).

For indoor conditions, the effective temperature (*ET*) is most commonly used. The effective temperature is the temperature at 50% relative humidity that yields the same total heat loss from the skin as for the actual environment (ASHRAE, 2005). For zero air velocity, the effective temperature is:

$$ET = t + w_{sk} i_m LR(P_w - 0.5\, P_{ws@ET}) \quad (10\text{-}1)$$

Where

ET = effective temperature, ℃

T = ambient temperature, ℃

w_{sk} = wetness of the skin, dimensionless

i_m = total vapor permeation efficiency, which is the ratio of actual latent to sensible heat transfer

LR = Luis Ratio, which is the ratio of evaporative heat transfer coefficient to convective heat transfer coefficient

P_w = partial pressure of water vapor

$P_{ws@ET}$ = saturated partial water vapor pressure at the effective temperature

Obviously, the total heat loss is also related to the ambient air velocity surrounding the skin. Furthermore, the heat transfer rate at the skin surface is also related to the radiant temperature, which may have a significant amount of heat exchange at the skin surface. Equation 10-1 and Figure 10-1 do not contain information on these two variables. For more detailed information about effective temperature, readers are referred to the *ASHRAE Fundamentals Handbook*, *Chapter 8: Thermal Comfort* (2005).

Table 10-1. **Typical indoor air contaminant threshold values for work places** (Zhang, 2005)

Contaminant	Symbol	Density (relative to air at room temperature)	Molar weight (g)		TWA	STEV	FEV
					(in parts per million [mg/L] for gases and mg/m^3 for dust)		
Gaseous (mg/L in Volume)							
Ammonia	NH_3	0.61	17	sharp, pungent	25	35	300[c]
Carbon dioxide	CO_2	1.57	44	none	5 000	30 000	
Carbon monoxide	CO	1.00	28	none	25	400[c]	
Formaldehyde	HCHO	1.06	30	pungent		0.3	
Hydroge n sulfide	H_2S	1.20	34	rotten eggs, nauseating	5	50	150
Methane	CH_4	0.57	16	none	1 000		50 000
Ozone	O_3	1.70	48	odorous	0.1		
Radon	Rn	7.84	222	none		4pCi/L[b]	
Particulate matter (mg/m^3)							
Animal building dust, inhalable					2.3[a]		
Animal building dust, respirable					0.23[a]		
Asbestos					0.1 fiber/mL		
Coal dust, respirable					0.4		
Flour dust, inhalable					0.5		
Grain dust, inhalable					4.0		
Iron oxide dust and fume, inhalable					5.0		
Metal dust, inhalable					10.0		
Welding fumes					5.0		

continued

Contaminant	Symbol	Density (relative to air at room temperature)	Molar weight (g)	TWA	STEV	FEV
				(in parts per million [mg/L] for gases and mg/m^3 for dust)		
Particulate not otherwise classified (PNOC), inhalable				10.0		
Particulate not otherwise classified (PNOC), respirable				3.0		

Source: If not specified, data are from: ACGIH, *Threshold limit values for chemical substances and physical agents and biological exposure indices*, American Conference of Governmental Industrial Hygienists: Cincinnati, OH. 2000.

[a] Donham, J. K., P. Haglind, Y. Peterson, R. Rylander, and L. Belin. *Environmental and health studies of workers in Swedish swine confinement buildings*, British Journal of Industrial Medicine, 40: 31-37, 1989.

[b] EPA, *Radon: Is your family at risk?* Environmental Protection Agency: Washington, DC. 1996.

[c] From Canada Plan Service, M-8710, 1985; M-9707, 1992. Ottawa, Canada. 1992.

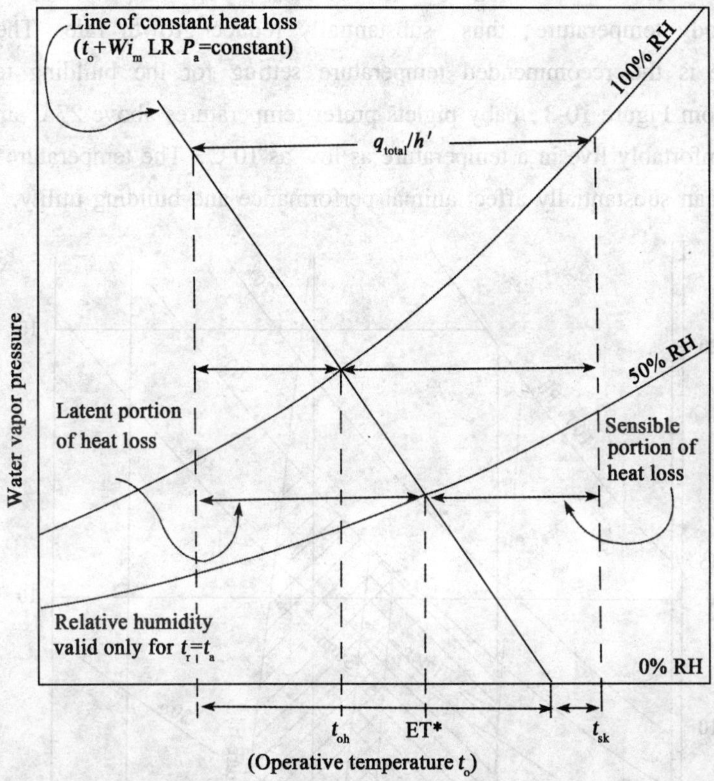

Figure 10-1. Constant heat loss line and its relationship to the effective temperature

A human comfort zone is shown in Figure 10-2. It is a very similar plot to Figure 2-3, but with the effective temperature in the picture.

10.2 Thermal Control for Livestock Facilities

Desirable environmental temperatures for different animal species are quite different. It is important that the environmental temperature is set for the needs of the specific animal species, rather than anthropomorphic perceptions. Desirable temperatures also vary substantially with the age or body mass of the same animal species. Figure 10-3 shows the temperature zones for swine buildings. The upper limit temperature refers to the temperature above which the pigs will suffer heat stress, thus, will substantially reduce feed intake and growth. Lower limit temperature refers to the temperature below which the pigs will primarily convert the feed intake to energy to maintain body temperature, thus, substantially reduce growth rate. The set-point temperature is the recommended temperature setting for the building temperature control. From Figure 10-3, baby piglets prefer temperatures above 27℃ and a 40 kg pig can comfortably live in a temperature as low as 10℃. The temperature setting for a building can substantially affect animal performance and building utility.

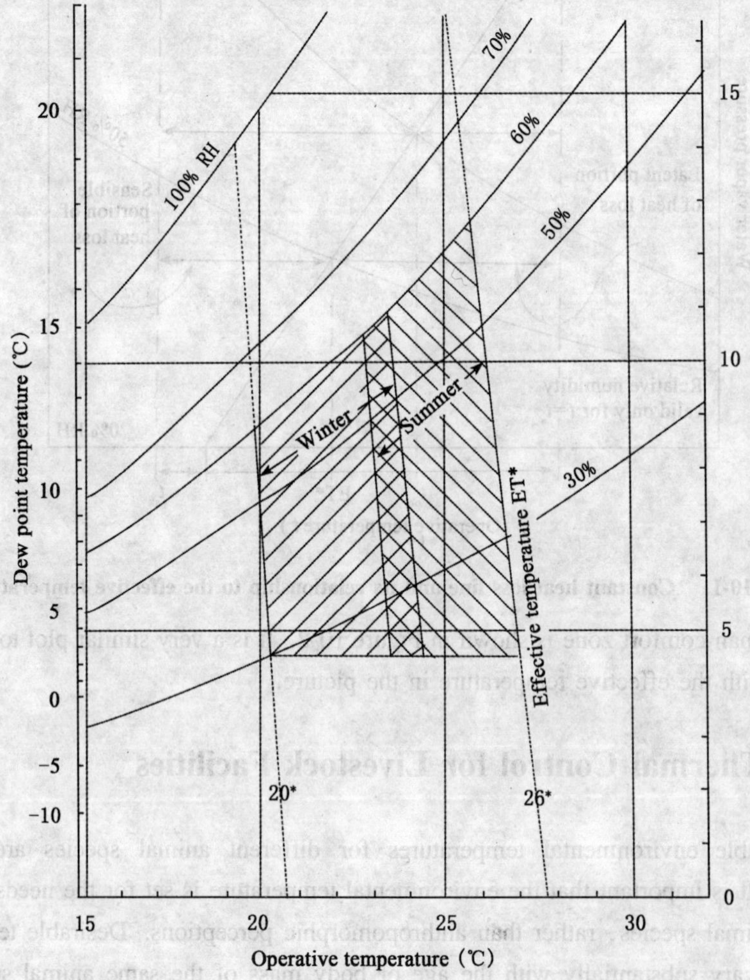

Figure 10-2. Thermal comfort zone for humans (ASHRAE, 2001)

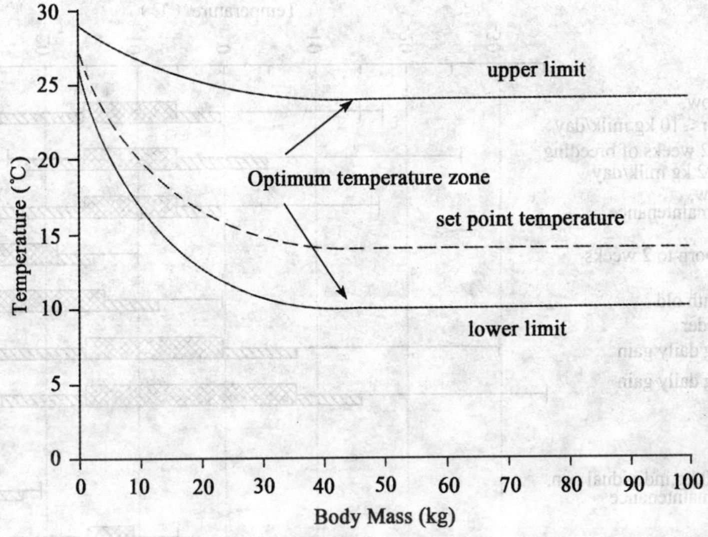

Figure 10-3. **Temperature control zone for pigs [adapted from the American Society of Agricultural Engineers (ASAE, 2000)]**

Temperature control for various livestock species are shown in Figure 10-4. It is noticeable that the controls for livestock are much broader than for humans, because the controls are based on the affect of the animal's performance (growth rate) rather than their comfort.

10.3 Heat Stress in Animal Production

Heat stress is a challenge for livestock throughout the world, even in some relatively cold climate regions. Technology exists to meet that challenge as long as livestock producers have made appropriate strategic decisions. In a hot climate, evaporative cooling can be effective for many animal species, including cattle and swine. Figure 10-5 shows the effect of ambient temperatures on swine metabolic energy intake. Metabolic intake represents the growth rate of the pigs. For example, for a 90 kg pig, every degree of ambient temperature increase higher than 20℃ will result in a three percent reduction in growth rate.

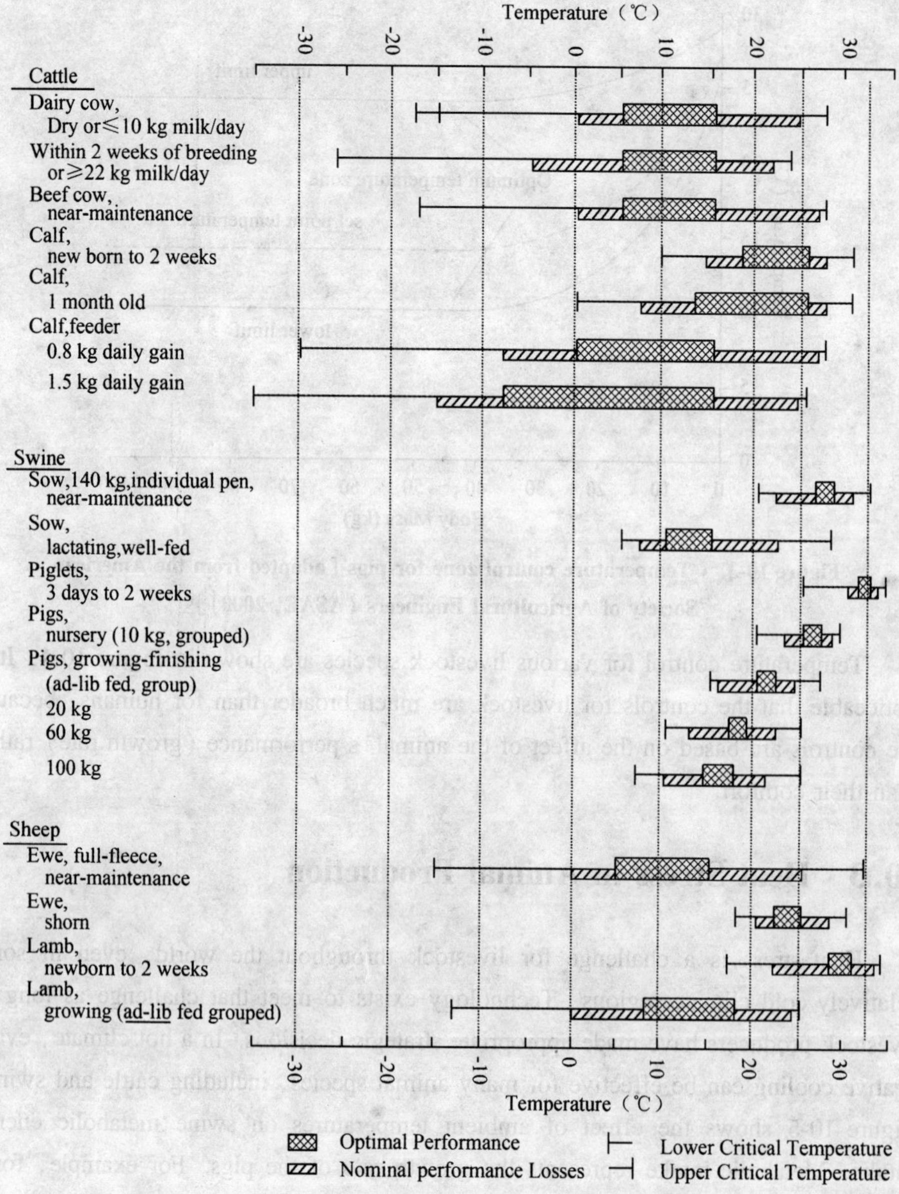

Figure 10-4. Temperature control limits for livestock (ASHRAE, 2007): (a) in SI units and (b) in English units

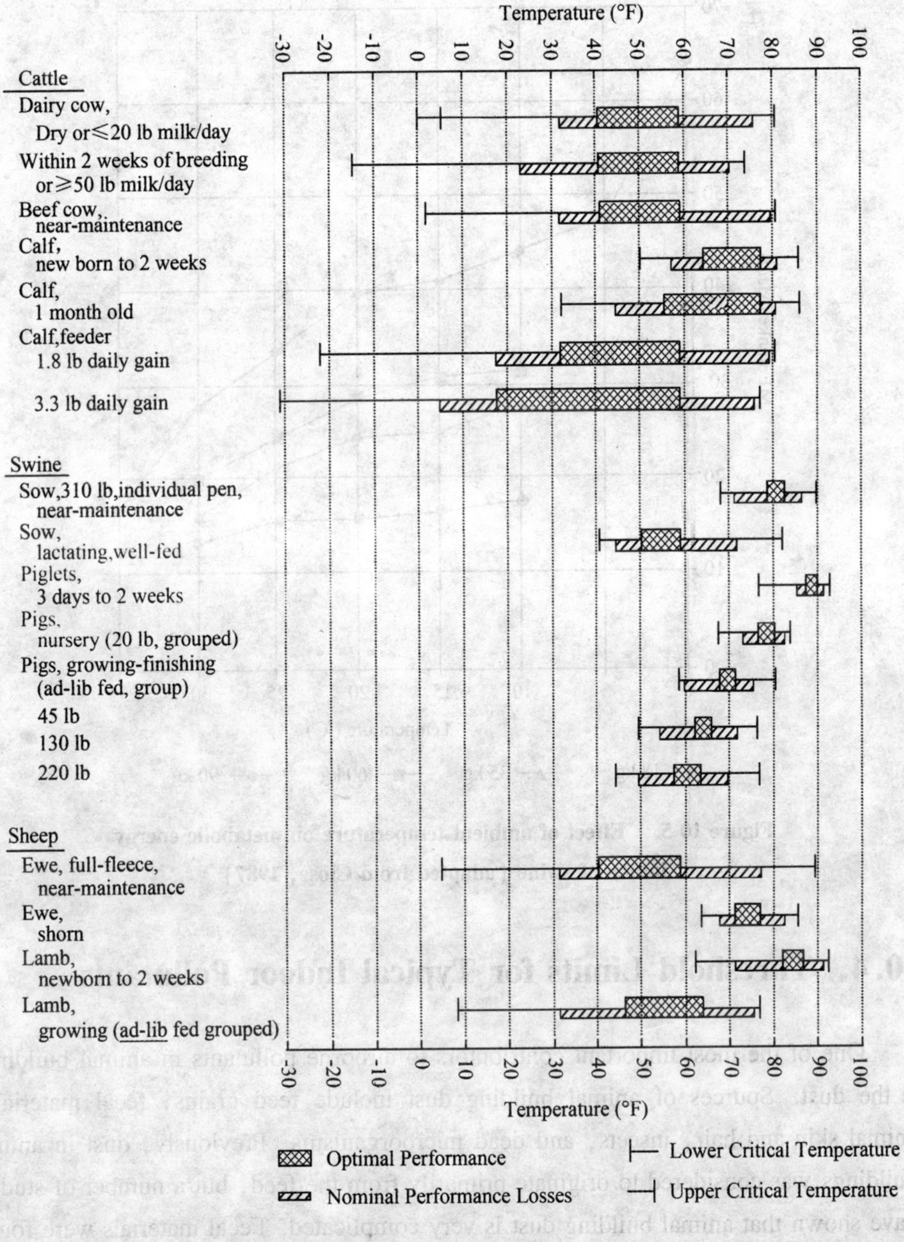

Figure 10-4. Continued

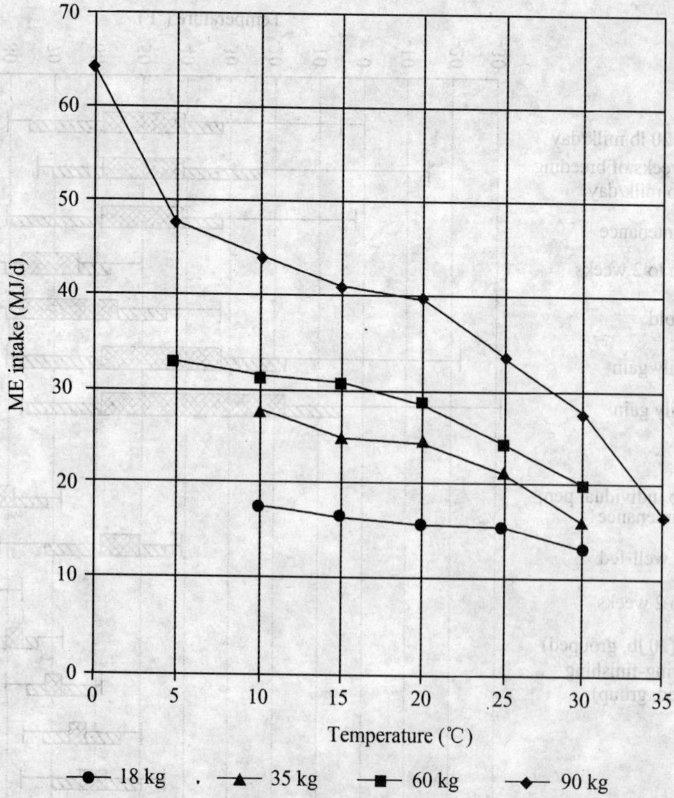

Figure 10-5. Effect of ambient temperature on metabolic energy intake of swine (adapted from Close, 1987)

10.4 Threshold Limits for Typical Indoor Pollutants

One of the most important contributors to airborne pollutants in animal buildings is the dust. Sources of animal building dust include feed grains, fecal materials, animal skin and hair, insects, and dead microorganisms. Previously, dust in animal buildings was considered to originate primarily from the feed, but a number of studies have shown that animal building dust is very complicated. Fecal materials were found to be a main constituent of airborne dust in turkey barns. In production animal buildings, especially with oil added to the feed, it is believed that fecal material and animals are the major dust sources.

Dust in animal buildings is biologically active and different from ordinary dust, such as field dust. Dust particles from animal buildings may combine components of viable organic compounds, mites, fungi, endotoxins, toxic gases, and other hazardous agents. Although the pathology of dusts in animal facilities remains unclear, there is little argument about their adverse affect on animal and human health. Some of the components, such as Streptococcus suis, are considered particularly hazardous to animals, and some such as the Alternaria species are suspected to be responsible for allergic symptoms in humans. In many animal production facilities, dusts tend to be accepted as a normal part of the job and may be ignored until permanent damage occurs. Toxic gases cause discomfort (e.g., nuisance odor or more dramatic symptoms), which attracts immediate attention and quick action in solving the problem. Dust control is one of the main tasks that will improve air quality in production animal facilities.

Household pets are one of the most common sources of allergic reactions to animals. Many people think that a pet allergy is caused by the fur of cats and dogs. However, researchers have found that the major allergens are proteins secreted by oil glands in the animals' skin and shed in dander as well as proteins in the saliva, which sticks to the fur when the animal licks itself. Urine is also a source of allergy-causing proteins. When the substance carrying the proteins dries, the proteins can then become airborne. Cats may be more likely than dogs to cause allergic reactions because they lick themselves more frequently, may be held more, spend more time indoors, and are closer to humans. Some rodents, such as guinea pigs and gerbils, have become increasingly popular as household pets. They, too, can cause allergic reactions in some people, as can mice and rats. Urine is the major source of allergens from these animals.

Allergies to animals can take two years or more to develop and may not subside until six months or more after ending contact with the animal. Carpet and furniture are a reservoir for pet allergens, and the allergens can remain in them for four to six weeks. In addition, these allergens can stay in the household air for months after the animal has been removed. Therefore, it is wise for people with an animal allergy to check with the landlord or previous owner to find out if furry pets have previously lived on the premises.

Threshold Limit Values (TLV's) refer to concentrations of airborne contaminants and represent conditions under which it is believed that nearly all workers may be repeatedly exposed day after day without adverse health effects. TLVs are based on available information from industrial experience, from experimental human and animal studies, and, when possible, from a combination of the three. The basis on which the values are established may differ from substance to substance; protection against impairment of health may be a guiding factor for some, whereas reasonable freedom from irritation, narcosis, nuisance, or other forms of stress may form the basis for others. Health impairments considered include those that shorten life expectancy, compromise physiological function, impair the capability for resisting other toxic substances or disease processes, or adversely affect reproductive faction or developmental processes.

Because of a wide variation in individual susceptibility, however, a small percentage may be affected more seriously by aggravation of a pre-existing condition or by development of an occupational illness. For example, smoking may act to enhance the biological effects of chemicals encountered in the workplace and may reduce the body's defense mechanisms against toxic substances. Individuals may be hyper susceptible or unusually responsive to some airborne contaminants because of genetic factors, age, personal habits (e.g., smoking, alcohol, or other drugs), medication, or previous exposures. Such workers may not be adequately protected from adverse effects from certain chemicals at concentrations at or below the threshold limits.

There are two categories of TLVs, differentiated by their legal status: one is mandatory and issued by governmental regulatory agencies such as the Environmental Protection Agency (EPA) and the Occupational Safety and Health Agency (OSHA) of the United States; the other is recommended and published by scientific communities such as the America Conference of Governmental and Industrial Hygienists (ACGIH) or individual researchers. The mandatory TLVs are enforced by law. The recommended TLVs are intended for use in the practice of industrial hygiene as guidelines or recommendations in the control of potential workplace health hazards. These limits are not fine lines between safe and dangerous concentrations, nor are they a relative index of toxicity. In spite of the fact that serious adverse health effects are not believed to occur as a result of exposure to the TLVs, the best practice is to maintain concentrations of all atmospheric contaminants as low as is practical.

The three categories of TLVs according to exposure are:

1. Threshold Limit Value—Time-Weighted Average (TLV-TWA) is the time-weighted average concentration for a conventional eight hour workday and a forty hour workweek, to which it is believed that nearly all workers may be repeatedly exposed, day-after-day, without adverse effect.

2. Threshold Limit Value—Short-Term Exposure Limit (TLV-STEL) is the concentration to which it is believed that workers can be exposed continuously for a short period of time without suffering from (1) irritation, (2) chronic or irreversible tissue damage, or (3) narcosis of sufficient degree to increase the likelihood of accidental injury, impair self-rescue or materially reduce work efficiency, provided that the daily TLV-TWA is not exceeded. It is not a separate independent exposure limit; rather, it supplements the time-weighted average (TWA) limit where there are recognized acute effects from a substance whose toxic effects are primarily of a chronic nature. STELs are recommended only where toxic effects have been reported from high short-term exposures in either humans or animals. A STEL is defined as a 15 minute TWA exposure that should not be exceeded at any time during a workday even if the eight hour TWA is within the TLV-TWA. Exposures above the TLV-TWA up to the STEL should be no longer than 15 minutes and should not occur more than four times per day. There should be at least 60 minutes between successive exposures in this range. An average period other than 15 minutes may be recommended when this is warranted by observed biological effects.

3. Threshold Limit Value - Ceiling (TLV-C) is the concentration that should not be exceeded during any part of the working exposure. An alternative term for TLV-C is fatal exposure values (FEV).

All three categories of TLVs are not available for every airborne contaminant. Many airborne contaminants have only one or two of the TLVs. It is important to observe that if any one of these types of TLVs is exceeded, a potential hazard from that contaminant is presumed to exist. Typical indoor air contaminants are listed in Table 10-1. More comprehensive lists of TLVs can be found from American Conference of Governmental Industrial Hygienists (ACGIH).

Whereas the ceiling limit places a definite boundary that concentrations should not be permitted to exceed, the TWA requires an explicit limit to the excursions that are permissible above the TLV-TWAs. TWAs permit excursions above the TLV provided they are compensated by equivalent excursions below the TLV-TWA during the workday. In some instances, it may be permissible to calculate the average concentration for a workweek rather than for a workday. The relationship between the TLV and permissible excursion is a rule of thumb and in certain cases may not apply. The amount by which the TLVs may be exceeded for short periods without injury to health depends upon a number of factors such as the nature of the contaminant, whether very high concentrations - even for short periods - produce acute poisoning, whether the effects are cumulative, the frequency with which high concentrations occur, and the duration of such periods. All factors must be taken into consideration in arriving at a decision as to whether a hazardous condition exists.

For the vast majority of substances with a TLV-TWA, there is not enough toxicological data available to warrant a STEL. Nevertheless, excursions above the TLV-TWA should be controlled even where the eight hour TLV-TWA is within recommended limits. While no definitive rationale was provided for these particular values, the basic concept was intuitive: in a well-controlled process exposure, excursions should be held within some reasonable limits. Unfortunately, neither toxicology nor collective industrial hygiene experience provide a solid basis for quantifying what those limits should be. For those air contaminants that have no STLV values, two simplified STEL criteria or excursion limit values are recommended by ACGIH:

1. Excursions in worker exposure levels may reach three times the TLV-TWA for no more than a total of 30 minutes during a workday; and

2. under no circumstances should they exceed five times the TLV-TWA.

The above simplification is based on the maximum recommended excursion observed in actual industrial processes. When the toxicological data for a specific substance are available to establish a STEL, this value takes precedence over the excursion limit regardless of whether it is more or less stringent.

Module 11: Room Air Cleaning Efficiency

The following discussion uses "particle" as an example pollutant in the air. Other particular pollutants can be treated in a similar fashion. Consider an enclosed airspace with an air cleaning device, such as a recirculation filter (Figure 11-1). The air filter has overall filtration efficiency, ξ_f, with a flow rate, Q_f, passing through the filter.

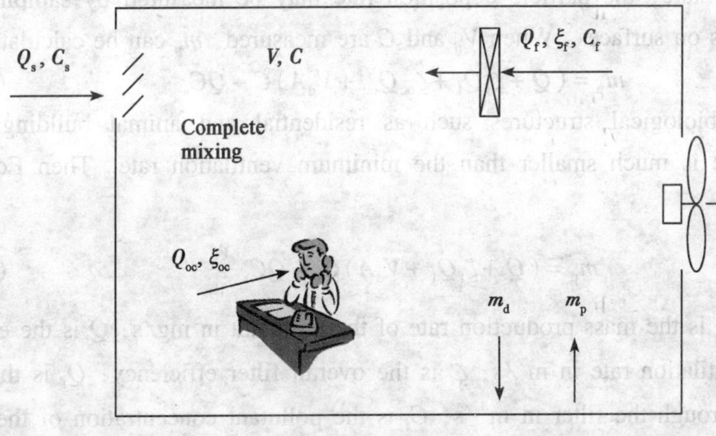

Figure 11-1. Variables and mass balance of airborne pollutants in a ventilated airspace

In order to derive a meaningful and solvable equation, the following assumptions are used. The ventilation rate, particle size distribution and thermal conditions within the room airspace do not vary with time. The airspace is completely mixed and there is no infiltration or exfiltration. When air in the airspace is completely mixed, the pollutant concentrations are the same for exhaust air and air within the airspace. The changes of total particle mass with time within the airspace can be expressed as

$$V\frac{\mathrm{d}C}{\mathrm{d}t} = m_p + Q_s C_s - Q_e C_e - \zeta_f Q_f C_f - \zeta_{oc} Q_{oc} C_{oc} - m_d \qquad (11\text{-}1)$$

Where V is the volume of the airspace in m^3; C_e, C_f and C_{oc} are particle mass concentrations in mg/m^3 for exhaust air, recirculation air and inspiration air of occupants, respectively; C_s is the particle concentration of the supply air in mg/m^3;

m_p is the production rate of particles in mg/s; Q_s and Q_e are air exchange rates for supply air and exhaust air in m^3/s, respectively; Q_f is the air flow rate passing a recirculation or an air filter in m^3/s; ζ_f is the particle removal efficiency of the recirculation or an air filter; Q_{oc} is the inspiration rate of the occupants within the airspace in m^3/s; ζ_{oc} is the particle removal efficiency of the respiration systems of the occupants; and m_d is the particle deposition rate in mg/s, and $m_d = V_d AC$, where V_d is the deposition velocity, in m/s and A is the deposition area, in m^2, and C is the pollutant concentration, in mg/m^3.

Note: The mass balance expressed by Equation 11-1 is valid for any other airborne contaminants such as gases.

At steady-state, the particle deposition rate may be measured by sampling the settled particles on surfaces. When V_d and C are measured, m_p can be calculated

$$m_p = (Q + \zeta_f Q_f + \zeta_{oc} Q_{oc} + V_d A) C - QC_s \tag{11-2}$$

In most biological structures such as residential and animal buildings, the inspiration rate is much smaller than the minimum ventilation rate. Then Equation 11-2 becomes

$$m_p = (Q + \zeta_f Q_f + V_d A) C_f - QC_s \frac{v_s}{v} \tag{11-3}$$

Where m_p is the mass production rate of the pollutant in mg/s; Q is the exhaust volumetric ventilation rate in m^3/s; ζ_f is the overall filter efficiency; Q_f is the flow rate passing through the filter in m^3/s; C_f is the pollutant concentration of the room with filter system in mg/m^3; C_s is the pollutant concentration of supply air in mg/m^3; v_s and v are specific volume of supply and room air in m^3/kg, respectively.

When the room has no filter, the pollutant production remains the same. The deposition velocity, V_d, pollutant concentration of the supply air, C_s, also remain the same as that in the room with a filter. The pollutant concentration within the airspace changes from C_f to C when there is no filter. The new mass balance is deduced from Equation 11-3 and becomes

$$m_p = (Q + V_d A) C - QC_s \frac{v_s}{v} \tag{11-4}$$

Because the pollutant production rates remain the same regardless of the filter, Equation 11-3 and Equation 11-4 are equal. The only change will be the pollutant concentration in the room. Equating Equation 11-3 and Equation 11-4 and rearranging, gives

$$\frac{C_f}{C}=\frac{Q+V_dA}{Q+\zeta_fQ_f+V_dA} \tag{11-5}$$

If we define the overall air cleaning efficiency of the room is the fraction of pollutant concentration reduction resulting from an air filtration within the room as ζ_r, then

$$\zeta_r=\frac{C-C_f}{C}=\frac{\zeta_fQ_f}{Q+\zeta_fQ_f+V_dA} \tag{11-6}$$

Note: The overall room air cleaning efficiency is always smaller than the filter efficiency. When the flow rate through the filter, Q_f, is substantially higher than the room ventilation rate, Q, the room cleaning efficiency is similar to the filter efficiency. In Equation 11-6, the deposition velocity, V_d, is usually an unknown variable and very difficult to measure. If the particle size distribution and the room ventilation systems are defined, V_d can be calculated. For gaseous pollutants, the deposition is negligible compared with the removal by ventilation. Thus, for gases,

$$\zeta_r=\frac{C-C_f}{C}=\frac{\zeta_fQ_f}{Q+\zeta_fQ_f} \tag{11-7}$$

From Equation 11-7, the overall room air cleaning efficiency is a function of filter efficiency flow rate through the filter and the room ventilation rate. If particles deposited on surfaces are much less than the particles removed by ventilation and filtration, Equation 11-7 can also be used to estimate the maximum particle removal efficiency for the room with a filter under complete mixing conditions. Figure 11-2 shows the relationship between the room air cleaning efficiency and the ratio of flow rate through the filter to the room ventilation rate (Q_f/Q). Clearly, at a given filter efficiency, ζ_f, the overall room air cleaning efficiency increases as the ratio Q_f/Q increases. On the other hand, increasing the flow rate through the filter means more frequent filter maintenance and operating cost.

Example 11-1

Determine the maximum room air cleaning efficiency for a room with a ventilation rate of 10 L/s. The room has an electrostatic precipitator (ESP) air cleaning device that has a filtration efficiency of 95% for particles and the flow rate passing through the filter is 0.5 L/s. Assume that deposition is negligible compared with the particle removal through ventilation. What is the room air cleaning efficiency? If you want the room air 50% cleaner, how many units of ESP air cleaner should be in the room?

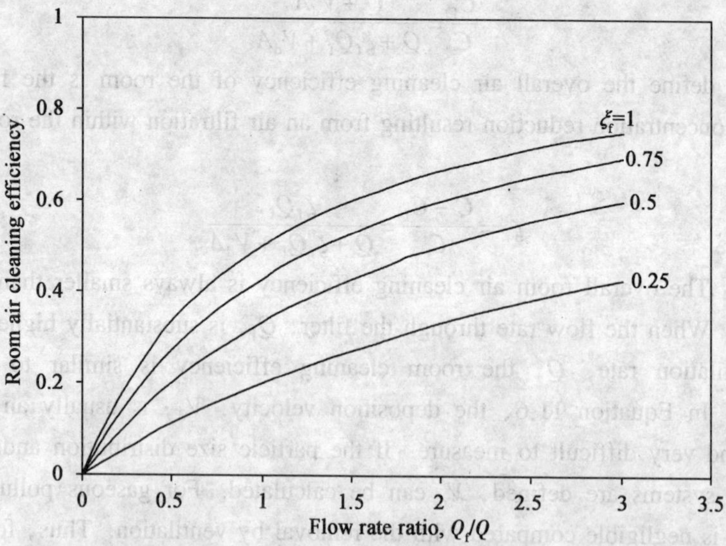

Figure 11-2. The overall efficiency for room air cleaning using filters at different filtration efficiencies

Solution

Neglecting the deposition for maximum possible cleaning efficiency,

$$\zeta_r = \frac{\zeta_f Q_f}{Q + \zeta_f Q_f} = \frac{0.95 \times 0.000\,5}{0.01 + 0.95 \times 0.000\,5} = 0.045\,3 = 4.53\%$$

For ζ_r = 50%, from Equation 11-7, the needed filtered air flow rate can be determined.

$$Q_f = \frac{Q\zeta_r}{\zeta_f(1-\zeta_r)} = \frac{0.01 \times 0.5}{0.95 \times (1-0.5)} = 0.010\,5(m^3/s) = 10.5(L/s)$$

Thus, 10.5 units of air cleaner each with a 0.5 L/s flow rate is needed.

11.1 Case Study: Evaluation of Air Cleaning Devices

11.1.1 Introduction

Indoor Air Quality is a hot topic in the residential consumer market these days. Millions of IAQ devices are sold each year and almost all claim to clean the air 100%

while virtually eliminating respiratory problems. The majority of consumers make buying decisions based on these claims. Do these products really stack up?

11.1.2 Objective

The objective of this case study is to use our engineering skills to determine how effective each product is at improving the quality of the air we breathe.

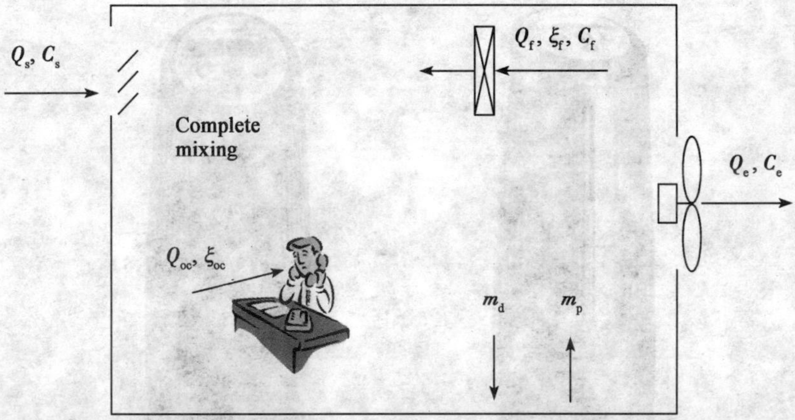

(1) Ozone Generator Air Cleaner

The ozone air cleaner claims to kill all bio-organisms (fungi, bacteria, etc.) and eliminate odors through the mystical powers of Ozone. In several university studies, such ozone generators were observed to produce 10 mg of ozone per hour. Calculate the following to see if this popular device really makes your air healthier!

a. If the generator is placed in a 500 ft^2 room with 8 ft ceilings and no ventilation, how long will it take for the room ozone concentration to reach the National Institute of Occupational Safety and Health (NIOSH) health limits of 0.10 mg/m^3 for Ozone?

b. (Hint: X ppm $= X$ $mg/m^3 \times 0.4731$ where X is the concentration of ozone.)

c. If 50 CFM supply and exhaust systems are added to the room what would the ozone concentration in the room be at steady-state? Is this value unsafe? (Assume the fresh air contains 0.05 mg/m^3 Ozone) CFM $= ft^3/min$.

d. Does this device work as advertised?

(2) Portable electrostatic precipitators (ESP)

These attractive sleek looking air cleaners are everywhere and often sell for over \$500. Boasting air cleaning efficiencies of 100% for particulate matter smaller than 10 microns (also known as PM_{10}), it's an easy sell, right? Calculate the following to make your buying decision:

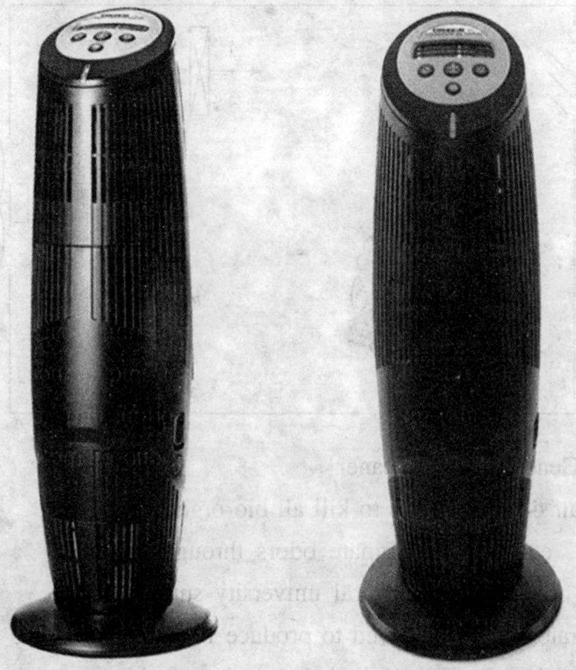

a. Ok, let's assume the manufacturers are honest people and we trust the 100% efficiency claim. Assuming a room with 50 CFM ventilation rate, what is the room cleaning efficiency of the ESP if it has a flow rate of 10 CFM?

b. If the outdoor concentration of PM_{10} is 35 $\mu g/m^3$ and there is a PM_{10} generation rate of 1 mg/h in the room, what is the steady state PM_{10} concentration in the room? Does this exceed the World Health Organizations PM_{10} limit of 20 $\mu g/m^3$ annual exposure?

c. Would you pay \$500 for this air cleaner? Why or why not?

(3) Ordinary house filter

Let's now look at the old school IAQ device, the \$10 house filter. This filter is very cheap and usually at least 80% efficient for PM_{10}. Install this filter in the room and see what happens. Assume this filter is placed in a recirculation system with a flow rate of 100 CFM. Assume the same room and outdoor conditions as the previous case. Find the following.

a. What is the room cleaning efficiency with the filter in the 100 CFM recirculation system? (Remember room ventilation rate is still 50 CFM as before.)

b. Assuming the conditions in part 2b, find the steady state PM_{10} concentration in the room. Does the new configuration meet the health standard?

c. If one filter lasts for 6 months, how long will you be able to clean your air with $500?

11.1.3 Discussion

- Which device would you rather use in your home? Why?
- Why do you feel IAQ devices are so profitable regardless of their effectiveness?
- What performance aspects are most important when choosing an indoor air cleaning device? (Stylish design, efficiency, flow rate, maintainability, etc.)

Module 12: Comprehensive Case Studies

12.1 Design a Cabin Minimum Air Flow Rate

12.1.1 Problem Description

You are working in the Research and Development (R/D) Department for the CDC Corporation. One day, CDC receives an order for an off-road service vehicle from the Department of Transportation of Alaska. The R/D chief engineer realizes that with the unique cold climate and harsh conditions, the existing CDC products need to be redesigned. He assigns you to design the ventilation system of the vehicle compartment. The following are some specifications of the vehicle:

- Battery operated (You must minimize the energy use!)
- Vehicle compartment is similar to a passenger car, 4.2 m^3 (2 m × 1.5 m × 1.4 m) in volume
- Total compartment surface area of 15 m^2
- Average insulation value of the compartment RSI = 1.2 $m^2 \cdot ℃/W$
- Maximum capacity: five people, including the driver

Inside conditions:

- Temperature, T = 20℃ (Assume at this temperature, approximately 50% of the total heat produced by people is sensible heat)
- Relative humidity (Φ) less than 70%
- Carbon dioxide concentration, C, is less than 3×10^{-3} m^3/m^3 air

Working conditions:

- Outside temperature (T_o) is −40℃
- Relative humidity (Φ_o) is 90%
- Carbon dioxide concentration (C_o) is 0.35×10^{-3} m^3/m^3 air

12.1.2 Find

The air flow rate for the first setting of the fan on the front panel (i.e., minimum ventilation rate) to satisfy the inside conditions: temperature, relative humidity and carbon dioxide concentration.

12.1.3 Solution

To meet the inside design conditions, you need to look at all three ventilation requirements for temperature, relative humidity and CO_2 concentration. Because the fresh air is supplied to the cabin by a supply fan and the cabin is pressurized, use the supply air (outside air), Q_o, to calculate the ventilation rate.

Step 1: Sensible heat balance (or temperature control) ventilation

From Equation 8-7, noting that the air to cabin is coming from the supply side (i.e., outside), then the volumetric flow rate for sensible heat balance is

$$Q_s = \frac{q_n v_s}{(h_{es} - h_{ss})}$$

The cabin air psychrometric properties can be found from a P-Chart:

$h_{es} = 20$ kJ/kg $\qquad h_{ss} = -40$ kJ/kg

$W_{es} = 10.2$ g/kg $\qquad W_{ss} = 0.1$ g/kg

$v_s = 0.66\ \text{m}^3/\text{kg} \qquad v_e = 0.842\ \text{m}^3/\text{kg}$

Further, the net sensible heat transfer rate of the cabin is

$$q_n = q_{as} + q_b + q_l + q_e + q_m + q_r$$

Where heat production from light, equipment and management are assumed negligible and from Table 7-1 for an average person, we assume the total heat production is 80 W/m^2 (between 65-115 W/m^2) for the driver and 60 W/m^2 for the four passengers. Further, from Table 7-2 we assume that 50% of the total heat is sensible heat and 50% is latent heat. The total heat and sensible heat produced by the people in the car are

$$THP = 80\ \frac{\text{W}}{\text{m}^2} \times 1.8\ \frac{\text{m}^2}{\text{person}} \times 1\ \text{person} + 60\ \frac{\text{W}}{\text{m}^2} \times 1.8\ \frac{\text{m}^2}{\text{person}} \times 4\ \text{person} = 576\ \text{W}$$

$$q_{as} = \frac{1}{2} THP = 288\ \text{W}$$

The total heat loss through the cabin shelter is

$$q_b = \frac{A}{RSI}(T_o - T) = \frac{15}{1.2} \times (-40 - 20) = -750\ \text{W}$$

" − " sign indicates a heat loss to the outside through the cabin shelter, then

$$q_n = q_{as} + q_b = 288\ \mathrm{W} + (-750\ \mathrm{W}) = -462\ \mathrm{W}$$

Substituting the above values into Equation 8-7, gives the ventilation required for sensible heat balance,

$$Q_s = \frac{q_n v_s}{(h_{es} - h_{ss})} = \frac{-0.462\left(\frac{\mathrm{kJ}}{\mathrm{s}}\right) \times 0.66\left(\frac{\mathrm{m^3}}{\mathrm{kg}}\right)}{20\left(\frac{\mathrm{kJ}}{\mathrm{kg}}\right) - \left(-40\frac{\mathrm{kJ}}{\mathrm{kg}}\right)} = -0.005\ 08\left(\frac{\mathrm{m^3}}{\mathrm{s}}\right)$$

The ventilation is negative. This means that in order to maintain temperature balance (cabin temperature at 20℃) it needs a negative ventilation, which is impractical.

Step 2: Moisture balance ventilation

Since 50% of the total heat produced by passengers is latent heat, i.e., q_w = 0.288 kW = 0.288 kJ/s, from Equation 7-2 we have the moisture production of the passengers

$$m_w = \frac{q_w}{2\ 501 + 1.805T} = \frac{0.288\frac{\mathrm{kJ}}{\mathrm{s}}}{(2\ 501 + 1.805 \times 20)\frac{\mathrm{kJ}}{\mathrm{kg}}} = 0.000\ 114\frac{\mathrm{kg}}{\mathrm{s}}$$

From Equation 8-10, the moisture balance ventilation is

$$Q_w = \frac{m_w v_s}{(W_{es} - W_{ss})} = \frac{0.000\ 114\frac{\mathrm{kg}}{\mathrm{s}} \times 0.66\frac{\mathrm{m^3}}{\mathrm{kg}}}{(0.010\ 2 - 0.000\ 1)\frac{\mathrm{kg}}{\mathrm{kg}}} = 0.007\ 58\frac{\mathrm{m^3}}{\mathrm{s}}$$

Step 3: CO_2 balance ventilation (exhaust).

The total volumetric CO_2 production can be calculated from the total heat production of the passengers

$$Q_{CO_2} = \frac{THP\left(\frac{\mathrm{kJ}}{\mathrm{s}}\right)}{24\ 600\frac{\mathrm{kJ}}{\mathrm{m^3}}} = \frac{0.576\frac{\mathrm{kJ}}{\mathrm{s}}}{24\ 600\frac{\mathrm{kJ}}{\mathrm{m^3}}} = 0.000\ 023\ 4\frac{\mathrm{m^3\,CO_2}}{\mathrm{s}}$$

From Equation 8-15,

$$Q_s = \frac{Q_{CO_2} v_s}{c_e v_e - c_o v_s} \times 10^6$$

$$= \frac{0.000\ 023\ 4\frac{\mathrm{m^3\,CO_2}}{\mathrm{s}} \times 0.66\frac{\mathrm{kg}}{\mathrm{m^3\,air}}}{3 \times 10^{-3}\frac{\mathrm{m^3}}{\mathrm{m^3\ air}} \times 0.842\frac{\mathrm{m^3}}{\mathrm{kg}} - 0.35 \times 10^{-3}\frac{\mathrm{m^3}}{\mathrm{m^3\ air}} \times 0.66\frac{\mathrm{m^3}}{\mathrm{kg}}} = 0.007\frac{\mathrm{m^3\,air}}{\mathrm{s}}$$

Comparing the three required ventilation rates, the minimum ventilation rate should be based on the moisture balance, i. e.,

$$Q_{min} = 0.00758 \text{ m}^3/\text{s}$$

12.1.4 Discussion Points

- If C_{CO_2} inside is 2×10^{-3} m³/m³ air, what should Q_{min} be?
- If only one person is inside, what will the inside temperature be?

12.2 Role Playing-Troubleshooting

12.2.1 Purpose

To troubleshoot a real world problem in a team environment using the knowledge acquired from this course.

12.2.2 Role Description

Bob: Experienced swine farmer and owner.

Cary: Young swine farm manager, Bob's Son.

Doug: Experienced engineer and owner of an engineering firm.

Pat: Young engineer-in-training working for DM Air Systems. He/she is a doer and the role involves some calculations.

12.2.3 Background and Scenario

The swine building described in Homework 4 is one of the barns owned by Mr. Robert (Bob) Doubtfire and his son, Cary. The building and ventilation system was designed by Doug McGyver, now a senior engineer and owner of the engineering firm—DM Air Systems. The barn was constructed eight years ago and has been operating satisfactorily. This winter, however, Doubtfire's barn had some problems with high ammonia concentrations. Ammonia concentrations in the barn on a cold day (at or below −20℃) reached 30 ppm. Due to the high ammonia levels, aggression in the pigs was high and some injuries occurred. Cary even had respiratory problems, which included watering eyes, coughing and a sore throat. Bob wants the problem solved as soon as possible.

Bob first called Doug and described the problem. Doug sent Pat, a young engineer-in-training, to the farm to investigate the problem and collect data for troubleshooting. Two days later, a meeting was arranged at Doug's firm to troubleshoot the problem.

12.2.4 Troubleshooting and Role Playing (your findings will be reported to the class later)

In the following session, your work team needs to assign the roles of: Bob, Cary, Doug, and Pat. If you have five team members, the fifth one can be a moderator and should be prepared to ask questions (you can challenge anyone in the play).

Each of you has your own work sheet provided by the director (instructor). Do not read any other role's manuscript. Take five minutes to digest your manuscript and be ready to play.

Five minutes later, four of you (plus the moderator, if any) get together and proceed as instructed in the manuscripts. Now …"**Action!**"

Bob's Manuscript

You are coming to the meeting to seek solutions to the problem. Since animal performance is down and employees are complaining about high ammonia, you are somewhat frustrated. The barn was designed by Doug's firm, so you feel that you have the right to challenge DM Air Systems. However, you know Doug is a reputable engineer and did an excellent job. You trust them for engineering solutions. You also believe that you have not made any mistakes in management which could cause the high ammonia production, but you are not sure.

Your major role in this meeting is as follows:

Describe briefly the problems you have. The building description is in Chapter 8 Problem. Some typical questions you may ask:

- I kept the minimum ventilation fans on all the time, why was the ammonia still so high?
- I have not had any moisture problem and the temperature control is excellent. How can I reduce the ammonia level immediately (short-term solutions)?
- How can I minimize the cost for long-term solutions (such as major

renovations) to this problem?

You may want to challenge the engineers with other questions.

Cary's Manuscript

As the son of one of the largest pork producers in the state, you decided to go into the pork production business. You help your dad with the daily management of the farm. You have most of the first hand information about the building and the animals. You also helped Pat collect some data when he/she visited your farm two days ago.

As manager of the building, you know of the following facts when the ammonia problem occurred:

1. Two of your employees were on sick leave (flu and allergy) for a week.
2. You only found one substitute to do the job.
3. During that period, you noticed that some animal pens were messy but you did not have enough employees to clean them.
4. You also noticed that the manure pit under the floor was very dry.

All of the above facts can cause more ammonia production within the building and, thus, the described problem. However, envisioning that this may still happen in the future you would like to have some methods to prevent the problem. Your objectives for coming to this meeting are to consult the engineers for solutions in emergency situations (short of employees) as well as longterm solutions.

Some of your example questions:

- When I enter the barn and find the ammonia level is too high, what can I do to deal with the problem?
- Can I open the doors to allow more ventilation? I sometimes do this for my house.
- What are the long-term solutions to this problem?

Doug's Manuscript

You are an experienced engineer and a man of integrity. You are very responsible to your clients. Two days ago you sent Pat, an assistant and an engineer-in-training, to investigate the problem. You know that the ventilation system is functioning and the minimum ventilation is measured at 2 000 L/s, just as designed. You try to identify the source of the problem and provide options for Bob. Here are some potential questions and comments you can use:

- Greetings to Bob and Cary (they are your customers, the meeting is in your firm, be a good host!)
- The minimum ventilation rate (2 000 L/s) was designed based on the recommendations for temperature and moisture balance control and it has been proven that it worked well for ammonia balance control in past years. Usually, such a minimum ventilation rate is sufficient for control of ammonia or other gases. It appears that something unusual occurred in that building.
- Ask Pat to describe his/her visit to the barn and the ammonia problem.
- Apparently more ventilation is needed under the unusual situation. Ask Pat (and work with him/her) on how much more ventilation is needed in addition to the 2 000 L/s.
- Bob and Cary, it appears that the ammonia production was unusually high. Did something unusual happen during that period? Such as dirty pens, delay of manure removal or dry manure pit?
- In answering Cary's question of "*Could I open the door for more ventilation*," you may suggest manually turning on the higher stage fans to provide more ventilation. Opening the door could chill young pigs (e.g., 25 kg body weight).
- Recommendations for a long-term solution:

◇ Install an additional small variable fan for additional minimum ventilation rate (1 000 L/s).

◇ The existing minimum ventilation fans are ten years old; it is time to replace them. Suggest to Bob that he should replace the existing minimum ventilation fans with slightly bigger variable fans that are capable of handling unusual situations. Use a ventilation graph.

Pat's Manuscript

You are a young, dynamic and enthusiastic engineer that has just joined the firm. You visited Bob's farm and collected the following data:

1. The animal building was set at 18℃ and the relative humidity was 60%.
2. On a cold day (−20℃ and 80% relative humidity) the ammonia

concentration reached as high as 0.03×10^{-3} m^3/m^3 air. Ammonia concentration outside is zero.

3. You also checked the fan specifications and measured the minimum ventilation rates of the fans of the building. The minimum fans deliver 2 000 L/s as designed.

Regardless of the high ammonia production, you need to provide more ventilation (add some fan capacity) to bring the ammonia down to a level lower than 0.02×10^{-3} m^3/m^3 air. You are Doug's assistant, and you are ready to work with Doug to solve Bob's problem.

Your potential role is:

Calculate the additional ventilation needed to bring the ammonia concentration to 0.02×10^{-3} m^3/m^3 air as follows: at 2 000 L/s ventilation, $C_{NH_3}=0.03\times10^{-3}$ m^3/m^3 air, you can calculate the ammonia production, V_p volume.

Since the ventilation balance is:

$$Q_e=\frac{m_c v_c v_e}{c_e v_e-c_o v_o}=\frac{Q_c v_e}{c_e v_e-c_o v_o}$$

Therefore, you can use $Q_e=2\ 000$ L/s to calculate Q_c, which is the ammonia volumetric production rate.

To deal with this unusually high ammonia production and keep the ammonia below 0.02×10^{-3} m^3/m^3 air, the new minimum ventilation rate should be:

Use the above equation again.

You can also indicate that to reduce the Q_c you can also reduce the ammonia concentration. Discuss with Bob and Cary better management practices such as more frequent pen cleaning, adding more water to the pit (which will suppress the ammonia release), and manually setting the fans at a higher rate temporarily.

12.3 Barn Description for Troubleshooting

12.3.1 Given

- A Central Illinois swine building is maintained at 18℃, less than or equal to 70% relative humidity, and less than or equal to 3×10^{-3} m^3/m^3 air in volume carbon dioxide.
- On a cold day, the outside temperature dropped to -20℃ and the relative

humidity is 80%.

- Deep floor temperature is approximately 10℃ year around.
- The building houses 1 200 pigs averaging 25 kg each.
- Carbon dioxide concentration of the outside air is 0.35×10^{-3} m^3/m^3 air in volume. The density of carbon dioxide is 1.53 times the air at the given temperature.
- Carbon dioxide production from animals can be calculated based on their total heat production because it is basically a metabolic product. On average, 1 L CO_2 for each 24.6 kJ of total heat produced.
- Building elevation is close to sea level.
- Building specifications are as following:

◇ Dimension: 50 m long, 12 m wide and 3 m high (floor to ceiling), rectangular.

◇ Insulation:

△ Walls with R20 insulation, $RSI = 3.94$ $m^2 \cdot$ ℃/W.

△ Ceiling with R30 insulation, $RSI = 5.48$ $m^2 \cdot$ ℃/W.

△ Concrete floor, $RSI = 1.76$ $m^2 \cdot$ ℃/W.

△ Concrete foundation with 50 mm polyethylene rigid insulation and 300 mm below exterior grade, $F = 0.4$ W/(m · ℃).

- Heat produced by other equipment, such as motors, is 500 W.
- Light: 30 lamps, 150 W each.
- Solar radiation is negligible in winter.

12.3.2 Useful information from a Ventilation graph

- Q_{min} and $Q_{max} = 30\ Q_{min}$.
- T_d (may be a moving target, depends on V min).
- Heater size (minimum).
- Not a single parameter (T, Φ, C) can control all weather conditions.

12.3.3 Troubleshooting Summary

Problem ID

- High NH_3 concentration

- Insufficient ventilation
- Health complaints
- Rusty equipment (worn out)
- Aggression

Analysis (reasoning, why)

- Dry manure
- Pigs are too crowded
- Dirty stalls
- Temperature? (outside)

Treatment

- Additional $V \approx 1\ 000$ L/s (variable fan)
- Better cleaning
- Replace a large variable fan
- Add water
- Change feed

Route

Problem ID ⟶Analysis ⟶Solution ⟶Evaluation

Problem ID

- High NH_3
- Health complaints
- Limited budget

Analysis (causes and problems)

- Unclean pens
- Outdated (insufficient) fans
- Pit was dry

Solutions

- Clean more frequently
- Upgrade the fans
- Ensure pit change
- Masks

Appendix 1: Building Sector Energy Consumption

The following summary data are from the **2008 U. S. Department of Energy Databook**, (http://buildingsdatabook. eere. enrgy. gov/). It contains energy consumption, expenditures, environmental impacts, and economic data related to the Buildings Sector. A new data section, Embodied Energy of Building Assemblies, contains data on energy used during the life-cycle of building materials. The following pieces of information give some insight into general trends in the Buildings Sector:

- Electricity energy consumption in the sector is increasing. Natural gas and petroleum energy consumption are declining.
- Less than 2% of annual Buildings Sector energy consumption is from renewable energy, each year from 1997 to 2030.
- Space heating is the largest energy end-use in the Buildings Sector. In 2006, it was 34% of site energy and 20% of primary energy.
- From 2006 to 2030, space heating, lighting, and space cooling are the top three energy end-uses (as a percentage of primary energy). Water heating and electronics are the next top two end-uses.
- Aggregate energy expenditures will have doubled from 1980 to 2030; the increase is 28% from 2006 to 2030.
- Electricity expenditures make up 67% of total Buildings Sector energy expenditures in 2006; in 2030, electricity expenditures are up to 70%.
- From 2006 to 2030, Buildings Sector electricity expenditures increase 34% to a total of $353 billion. Natural gas increases 19% to a total of $112 billion. Petroleum increases 5.5% to a total of $36 billion.
- The average price of electricity in the Residential Sector in 2006 was 10.4 cents per kW · h; 9.5 cents per kW · h for the Commercial Sector.
- Space heating, lighting, and space cooling are the top three energy end-use expenditures.
- New buildings construction is $785 billion in 2006. Building improvements and repairs is $438 billion.
- Carbon dioxide emissions by utilities generating, transmitting, and distributing electricity drives the Buildings Sector carbon dioxide emissions.

- The Buildings Sector percentage of carbon dioxide emissions increases from 38% in 2006 to 43% in 2030. Emissions in 1980 were 33%.
- World carbon dioxide emissions increased 1.9% per year from 1990 to 2005. Emissions are projected to increase 2.1% per year from 2005 to 2010.

Buildings Energy Data Book: **1.1 Buildings Sector Energy Consumption** September 2008

1.1.1 U.S. Residential and Commercial Buildings Total Primary Energy Consumption (Quadrillion Btu and Percent of Total)

	Natural Gas		Petroleum (1)		Coal		Renewable (2)		Electricity				TOTAL (2)		Growth Rate
									Sales	Losses	Total				2006 - Year
1980	7.52	28%	3.04	11%	0.15	0.6%	0.87	3.3%	4.35	10.51	14.86	56.2%	26.43	100%	-
1990	7.22	24%	2.36	8%	0.15	0.5%	0.74	2.4%	6.01	13.92	19.93	65.6%	30.41	100%	-
2000	8.35	22%	2.32	6%	0.10	0.3%	0.63	1.7%	8.02	18.26	26.28	69.8%	37.68	100%	-
2006	**7.42**	**19%**	**1.93**	**5%**	**0.09**	**0.2%**	**0.58**	**1.5%**	**9.05**	**19.70(3)**	**28.75**	**74.2%**	**38.77**	**100%**	-
2010	7.99	19%	1.95	5%	0.09	0.2%	0.62	1.5%	9.67	20.71	30.38	74.0%	41.04	100%	1.4%
2015	8.46	20%	2.00	5%	0.09	0.2%	0.61	1.4%	10.22	21.59	31.81	74.0%	42.97	100%	1.1%
2020	8.77	19%	2.01	4%	0.09	0.2%	0.61	1.3%	10.92	23.04	33.96	74.7%	45.45	100%	1.1%
2025	8.98	19%	1.99	4%	0.09	0.2%	0.61	1.3%	11.68	24.44	36.11	75.6%	47.78	100%	1.1%
2030	9.11	18%	1.97	4%	0.09	0.2%	0.61	1.2%	12.50	25.82	38.32	76.5%	50.10	100%	1.1%

Note(s): 1) Petroleum Includes distilate and residual fuels, liquefied petroleum gas, kerosene, and motor gasoline. 2) Includes *site* - marketed and non - marketed renewable energy. 3) 2006 *site* - to - source electricity conversion = 3.18.

Source(s): EIA. State Energy Data 2005: Consumption, February 2008, Tables 8 - 12. p. 18 - 22 for 1980 - 2005; and EIA, Annual Energy Outlook 2008, Mar, 2008, Table A2, p. 117 - 119 for 2006 - 2030 and Table A17, p. 143 - 144 for non - marketed renewable energy.

1.1.2 U.S. Buildings *Site* Renewable Energy Consumption (Quadrillion Btu) (1)

	Wood(2)	Solar Thermal(3)	Solar PV(3)	GSHP(4)	Total	Growth Rate 2006 - Year
1980	0.858	0.000	0.000	0.000	0.858	-
1990	0.609	0.056	0.000	0.003	0.668	-
2000	0.559	0.024	0.000	0.017	0.599	-
2006	**0.538**	**0.038**	**0.001**	**0.003**	**0.581**	-
2010	0.570	0.043	0.004	0.004	0.621	1.7%
2015	0.547	0.052	0.004	0.006	0.609	0.5%
2020	0.533	0.059	0.006	0.008	0.607	0.3%
2025	0.520	0.066	0.010	0.011	0.607	0.2%
2030	0.508	0.073	0.016	0.014	0.611	0.2%

Note(s): 1) Does not include renewable energy consumed by electric utilities (including hydroelectric). 2) Includes wood and wood waste, municipal solid waste, and other biomass used by the commercial sector to cogenerate electricity. 3) Includes only solar energy. 4) GHP = Ground - coupled heat pumps.

Source(s): EIA, State Energy Data 2005: Consumption, February 2008, Tables 8 - 12, p. 18 - 22 for 1980 - 2000; and EIA, Annual Energy Outlook 2008, Mar. 2008, Table A17, p. 143 - 144 for 2006 - 2030; Annual Energy Outlook 2006, Feb. 2006, Table A17, p. 159; EIA, Annual Energy Outlook 2005, Jan. 2005, Table A17, p. 163; EIA, Annual Energy Outlook 2004, Jan. 2004, Table A17, p, 157; EIA, Annual Energy Outlook 2002, Dec. 2001, Table A18, p. 148 For 1999 - 2004 Solar

1.1.3 Building Share of U.S. Primary Energy Consumption (Percent)

	Buildings						Total Consumption
	Residential	Commercial	Total	Industry	Transportation	Total	(quads)
1980(1)	27.4%	18.3%	**45.7%**	36.0%	18.3%	100%	57.9
1990	22.4%	17.5%	**40.0%**	38.9%	21.1%	100%	76.1
2000	21.1%	17.7%	**38.8%**	36.1%	25.2%	100%	97.2
2006	**20.9%**	**18.0%**	**38.9%**	**32.7%**	**28.4%**	**100%**	**99.5**
2010	21.5%	18.1%	**39.7%**	32.2%	28.1%	100%	103.3
2015	21.0%	19.0%	**40.0%**	31.6%	28.4%	100%	107.3
2020	21.1%	19.8%	**40.9%**	30.9%	28.2%	100%	110.8
2025	21.1%	20.6%	**41.6%**	30.5%	27.9%	100%	114.5
2030	21.2%	21.2%	**42.4%**	29.6%	28.0%	100%	118.0

Note(s): 1) Renewables are not included in the 1980 data.

Source(s): EIA, State Energy Data 2005; Consumption, February 2008, Tables 8 – 12, p. 18 – 22 for 1980 – 2005; and EIA, Annual Energy Outlook 2008, Mar. 2008, Table A2, p. 117 – 119 for 2006 – 2030 data and Table A17, p. 143 – 144 for non – marketed renewable energy.

Buildings Energy Data Book: **1.1 Buildings Sector Energy Consumption** September 2008

1.1.4 2006 U.S. Buildings Energy End-Use Splits, by Fuel Type (Quadrillion Btu)

	Natural Gas	Fuel Oil (1)	LPG	Other Fuel(2)	Renw. En. (3)	Site Electric	Site Total	Site Percent	Primary Electric (4)	Primary Total	Primary Percent
Space Heating (5)	4.31	0.84	0.23	0.18	0.41	0.53	6.50	34.1%	1.69	7.66	19.8%
Lighting						2.16	2.16	11.3%	6.86	6.86	17.7%
Space Cooling	0.02					1.54	1.56	8.2%	4.89	4.91	12.7%
Water Heating	1.63	0.15	0.06		0.04	0.58	2.45	12.9%	1.85	3.72	9.6%
Electronics (6)						0.96	0.96	5.0%	3.04	3.04	7.8%
Refrigeration (7)						0.70	0.70	3.7%	2.23	2.23	5.8%
Cooking	0.45		0.03			0.27	0.75	3.9%	0.85	1.33	3.4%
Wet Clean (8)	0.07					0.38	0.46	2.4%	1.22	1.30	3.3%
Ventilation (9)						0.35	0.35	1.8%	1.10	1.10	2.8%
Computers						0.28	0.28	1.5%	0.89	0.89	2.3%
Other (10)	0.27	0.02	0.23	0.05	0.13	0.82	1.52	8.0%	2.60	3.30	8.5%
Adjust to SEDS (11)	0.67	0.23				0.48	1.37	7.2%	1.54	2.43	6.3%
Total	**7.42**	**1.24**	**0.55**	**0.23**	**0.58**	**9.05**	**19.06**	100%	**28.75**	**38.77**	100%

Note(s): 1) Includes distillate fuel oil (1.12 quad) and residual fuel oil (0.9 quad). 2) Kerosene (0.12 quad) and coal (0.09 quad) are assumed attributable to space heating. Motor gasoline (0.05 quad) assumed attributable to other end-uses. 3) Comprised of wood space heating (0.41 quad), biomass (0.13 quad), solar water heating

(0.03 quad), geothermal space heating (less than 0.01 quad), and solar photovoltaics (PV) less than 0.01 quad. 4) Site-to-source electricity conversion (due to generation and transmission losses) = 3.18. 5) Includes furnace fans (0.21 quad). 6) Includes color television (1.05 quad) and other office equipment (0.64 quad). 7) Includes refrigerators (1.24 quad) and freezers (0.49 quad). Includes commercial refrigeration. 9) Includes clothes washers (0.11 quad), natural gas clothes dryers (0.07 quad), electric clothes dryers (0.81 quad) and dishwashers (0.3 quad). Does not include water heating energy. 8) Commercial only; residential fan and pump energy use included proportionately in space heating and cooling. 10) Includes residential small electric devices, heating elements, motors, swimming pool heaters, hot tub heaters, outdoor grills, and natural gas outdoor lighting. Includes commercial service station equipment, ATMs, telecommunications equipment, medical equipment, pumps, emergency electric generators, combined heat and power in commercial buildings, and manufacturing performed in commercial buildings. 11) Energy adjustment EIA uses to relieve discrepancies between data sources. Energy attributable to the residential and commercial buildings sector, but not directly to specific end-uses.

Source(s): EIA, Annual Energy Outlook 2008, Mar. 2008, Tables A2, p. 117-119, Table A4, p. 122-123, Table A5, p. 124-125, and Table A17, p. 143-144; EIA, National Energy Modeling System (NEMS) for AEO 2008, Mar. 2008; BTS/A. D. Little, Electricity Consumption by Small End-Uses in Residential Buildings, Aug. 1998, Appendix A for residential electric end-uses; BTS/A. D. Little, Energy Consumption Characteristics of Commercial Building HVAC Systems, Volume II: Thermal Distribution, Auxiliary Equipment, and Ventilation, Oct. 1999, p. 11-2 and 5-25-5:26; EIA, Annual Energy Outlook 1998, Dec. 1997, Table A5, p. 108-109 ventilation; BTP/Navigant Consulting, U.S. Lighting Market Characterization, Volume I, Sept. 202, Table 8-2, p. 63; and EIA, Supplement to the AEO 2008, April 2008, Table 22.

2006 U.S. Buildings Energy End-Use Splits

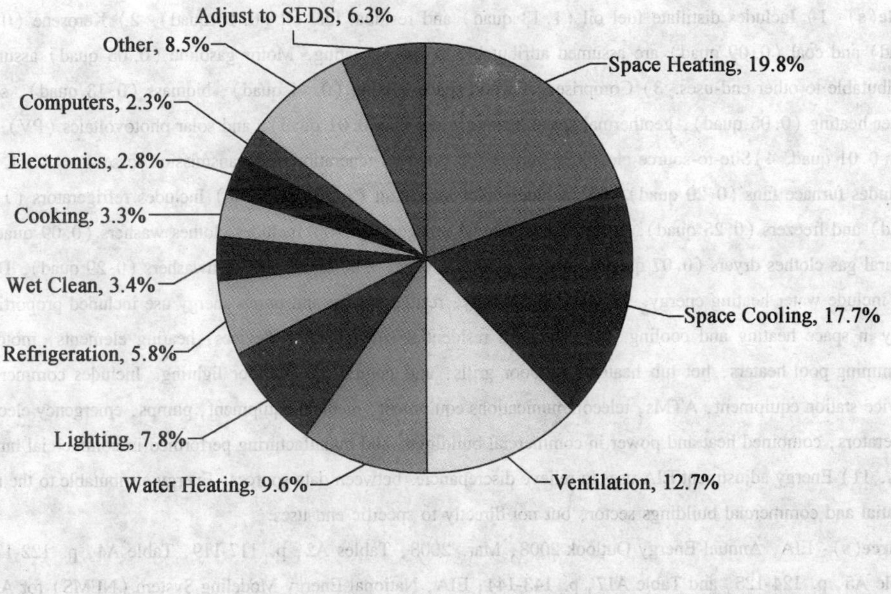

Buildings Energy Data Book: **1.1 Buildings Sector Energy Consumption** September 2008

1.1.5 2010 U.S. Buildings Energy End-Use Splits, by Fuel Type (Quadrillion Btu)

	Natural Gas	Fuel Oil (1)	LPG	Other Fuel(2)	Renw. En. (3)	Site Electric	Site Total	Site Percent	Primary Electric (4)	Primary Total	Primary Percent
Space Heating (5)	4.86	0.89	0.24	0.19	0.44	0.50	7.13	35.1%	1.59	8.21	20.0%
Lighting						1.29	1.29	6.3%	5.78	5.78	14.1%
Space Cooling	0.02					0.19	0.21	1.0%	4.04	4.06	9.9%
Water Heating	1.62	0.14	0.05		0.04	0.54	2.39	11.7%	1.69	3.54	8.6%
Electronics (6)						1.84	1.84	9.0%	2.96	2.96	7.2%
Refrigeration (7)						0.68	0.68	3.4%	2.14	2.14	5.2%
Wet Clean (8)	0.07					0.94	1.02	5.0%	1.19	1.27	3.1%
Computers						0.35	0.35	1.7%	1.10	1.10	2.7%
Cooking	0.47		0.03			0.38	0.88	4.3%	0.46	0.96	2.3%
Ventilation (9)						0.15	0.15	0.7%	0.60	0.60	1.5%
Other (10)	0.29	0.02	0.25	0.05	0.13	2.02	2.76	13.6%	6.35	7.09	17.3%
Adjust to SEDS (11)	0.66	0.19				0.80	1.64	8.1%	2.50	3.34	8.1%
Total	**7.99**	**1.23**	**0.57**	**0.24**	**0.62**	**9.67**	**20.33**	100%	**30.38**	**41.04**	100%

Note(s): 1) Includes distillate fuel oil (1.13 quad) and residual fuel oil (0.10 quad). 2) Kerosene (0.08 quad) and coal (0.09 quad) are assumed attributable to space heating. Motor gasoline (0.05 quad) assumed attributable to other end-uses. 3) Comprised of wood space heating (0.44 quad), biomass (0.13 quad), solar water heating (0.05 quad), geothermal space heating (less than 0.01 quad), and solar photovoltaics (PV) less than 0.01 quad. 4) Site-to-source electricity conversion (due to generation and transmission losses) = 3.14. 5) Includes furnace fans (0.20 quad). 6) Includes color television (1.23 quad). 7) Includes refrigerators (1.89 quad) and freezers (0.25 quad). Includes commercial refrigeration. 8) Includes clothes washers (0.09 quad), natural gas clothes dryers (0.07 quad), electric clothes dryers (0.80 quad) and dishwashers (0.29 quad). Does not include water heating energy. 9) Commercial only; residential fan and pump energy use included proportionately in space heating and cooling. 10) Includes residential small electric devices, heating elements, motors, swimming pool heaters, hot tub heaters, outdoor grills, and natural gas outdoor lighting. Includes commercial service station equipment, ATMs, telecommunications equipment, medical equipment, pumps, emergency electric generators, combined heat and power in commercial buildings, and manufacturing performed in commercial buildings. 11) Energy adjustment EIA uses to relieve discrepancies between data sources. Energy attributable to the residential and commercial buildings sector, but not directly to specific end-uses.

Source(s): EIA, Annual Energy Outlook 2008, Mar. 2008, Tables A2, p. 117-119, Table A4, p. 122-123, Table A5, p. 124-125, and Table A17, p. 143-144; EIA, National Energy Modeling System (NEMS) for AEO 2008, Mar. 2008; and EIA, Supplement to the AEO 2008, April 2008, Table 22.

2010 U.S. Buildings Energy End-Use Splits

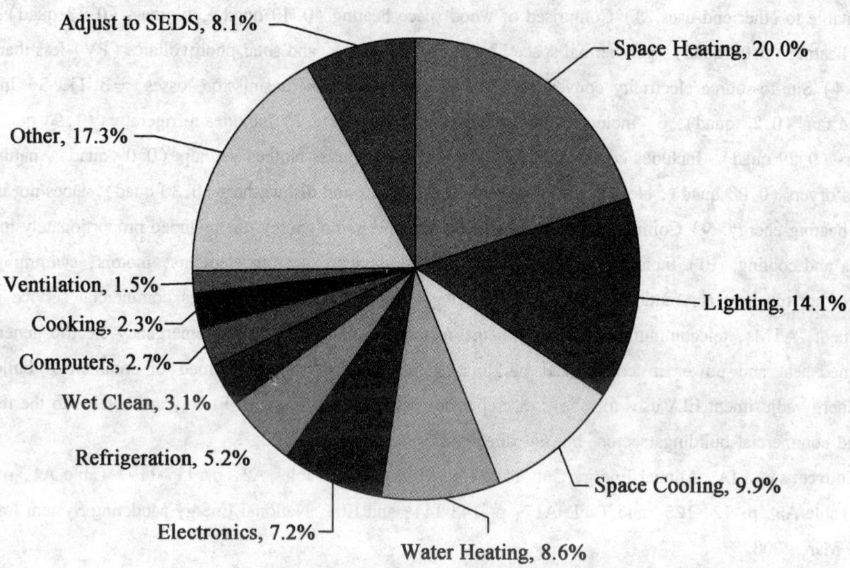

Buildings Energy Data Book: **1.1 Buildings Sector Energy Consumption** September 2008

1.1.6 2020 U.S. Buildings Energy End-Use Splits, by Fuel Type (Quadrillion Btu)

	Natural Gas	Fuel Oil (1)	LPG	Other Fuel(2)	Renw. En. (3)	Site Electric	Site Total	Site Percent	Primary Electric (4)	Primary Total	Primary Percent
Space Heating (5)	5.23	0.90	0.24	0.19	0.41	0.54	7.51	33.5%	1.68	8.65	19.0%
Lighting						1.73	1.73	7.7%	5.37	5.37	11.8%
Space Cooling	0.02					1.46	1.48	6.6%	4.53	4.55	10.0%
Water Heating	1.80	0.13	0.05		0.06	0.58	2.62	11.7%	1.81	3.85	8.5%
Electronics (6)						1.22	1.22	5.4%	3.79	3.79	8.3%
Refrigeration (7)						0.71	0.71	3.2%	2.21	2.21	4.9%
Computers						0.42	0.42	1.9%	1.31	1.31	2.9%
Wet Clean (8)	0.08					0.39	0.47	2.1%	1.22	1.30	2.9%
Cooking	0.54		0.03			0.16	0.73	3.3%	0.50	1.08	2.4%
Ventilation (9)						0.21	0.21	0.9%	0.65	0.65	1.4%
Other (10)	0.38	0.02	0.30	0.05	0.14	2.62	3.51	15.6%	8.14	9.03	19.9%
Adjust to SEDS (11)	0.72	0.19				0.88	1.80	8.0%	2.74	3.66	8.1%
Total	**8.77**	**1.25**	**0.61**	**0.25**	**0.61**	**10.92**	**22.41**	100%	**33.96**	**45.45**	100%

Note(s): 1) Includes distillate fuel oil (1.14 quad) and residual fuel oil (0.10 quad). 2) Kerosene (0.08 quad) and coal (0.09 quad) are assumed attributable to space heating. Motor gasoline (0.05 quad) assumed attributable to other end-uses. 3) Comprised of wood space heating (0.40 quad), biomass (0.13 quad), solar water heating (0.06 quad), geothermal space heating (0.01 quad), and solar photovoltaics (PV) less than 0.01 quad. 4) Site-to-source electricity conversion (due to generation and transmission losses) = 3.11. 5) Includes furnace fans (0.23 quad). 6) Includes color television (1.33 quad). 7) Includes refrigerators (1.93 quad) and freezers (0.29 quad). Includes commercial refrigeration. 8) Includes clothes washers (0.09 quad), natural gas clothes dryers (0.08 quad), electric clothes dryers (0.84 quad) and dishwashers (0.30 quad). Does not include water heating energy. 9) Commercial only; residential fan and pump energy use included proportionately in space heating and cooling. 10) Includes residential small electric devices, heating elements, motors, swimming pool heaters, hot tub heaters, outdoor grills, and natural gas outdoor lighting. Includes commercial service station equipment, ATMs, telecommunications equipment, medical equipment, pumps, emergency electric generators, combined heat and power in commercial buildings, and manufacturing performed in commercial buildings. 11) Energy adjustment EIA uses to relieve discrepancies between data sources. Energy attributable to the residential and commercial buildings sector, but not directly to specific end-uses.

Source(s): EIA, Annual Energy Outlook 2008, Mar. 2008, Tables A2, p. 117-119, Table A4, p. 122-123, Table A5, p. 124-125, and Table A17, p. 143-144; and EIA, National Energy Modeling System for AEO 2008, Mar. 2008.

2020 U.S. Buildings Energy End-Use Splits, by Fuel Type

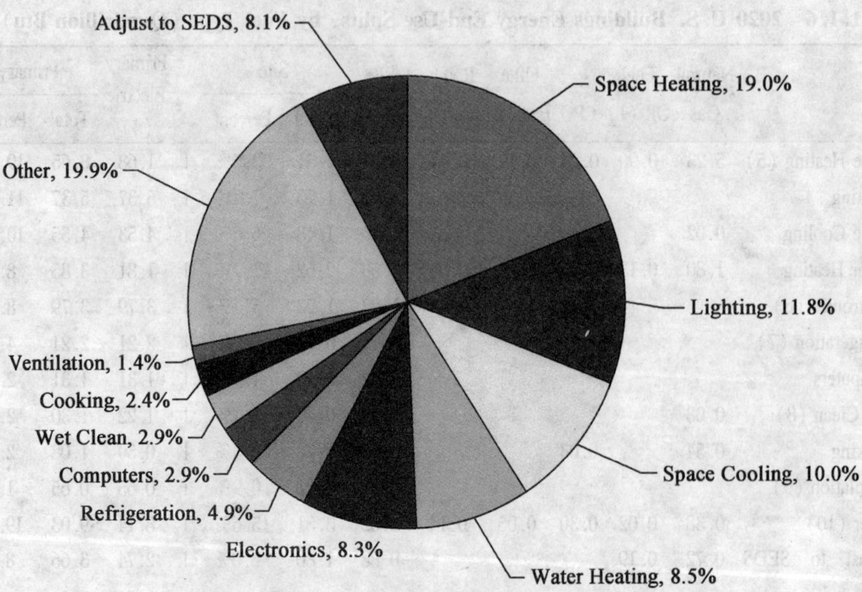

Buildings Energy Data Book: **1.1 Buildings Sector Energy Consumption** September 2008

1.1.7 2030 U.S. Buildings Energy End-Use Splits, by Fuel Type (Quadrillion Btu)

	Natural Gas	Fuel Oil (1)	LPG	Other Fuel(2)	Renw. En. (3)	Site Electric	Site Total	Site Percent	Primary Electric (4)	Primary Total	Primary Percent
Space Heating (5)	5.30	0.84	0.23	0.19	0.39	0.56	7.51	30.9%	1.71	8.67	17.3%
Lighting						1.83	1.83	7.5%	5.61	5.61	11.2%
Space Cooling	0.02					1.65	1.67	6.9%	5.06	5.08	10.1%
Water Heating	1.82	0.12	0.04		0.07	0.59	2.65	10.9%	1.81	3.87	7.7%
Electronics (6)						1.47	1.47	6.0%	4.50	4.50	9.0%
Refrigeration (7)						0.78	0.78	3.2%	2.40	2.40	4.8%
Computers						0.51	0.51	2.1%	1.56	1.56	3.1%
Wet Clean (8)	0.08					0.43	0.51	2.1%	1.31	1.40	2.8%
Cooking	0.59		0.03			0.17	0.80	3.3%	0.54	1.16	2.3%
Ventilation (9)						0.23	0.23	1.0%	0.71	0.71	1.4%
Other (10)	0.62	0.02	0.34	0.05	0.15	3.30	4.47	18.4%	10.11	11.28	22.5%
Adjust to SEDS (11)	0.67	0.19				0.97	1.84	7.6%	2.99	3.85	7.7%
Total	**9.11**	**1.17**	**0.64**	**0.25**	**0.61**	**12.50**	**24.28**	100%	**38.32**	**50.09**	100%

Note(s): 1) Includes distillate fuel oil (1.45 quad) and residual fuel oil (0.12 quad). 2) Kerosene (0.11 quad) and coal (0.10 quad) are assumed attributable to space heating. Motor gasoline (0.05 quad) assumed attributable to other end-uses. 3) Comprised of wood space heating (0.38 quad), biomass (0.13 quad), solar water heating (0.07 quad), geothermal space heating (less than 0.01 quad), and solar photovoltaics (PV) 0.02 quad. 4) Site-to-source electricity conversion (due to generation and transmission losses) = 3.07. 5) Includes furnace fans (0.25 quad). 6) Includes color television (1.69 quad) and other office equipment (2.81 quad). 7) Includes refrigerators (2.10 quad) and freezers (0.34 quad). Includes commercial refrigeration. 8) Includes clothes washers (0.08 quad), natural gas clothes dryers (0.08 quad), electric clothes dryers (0.91 quad) and dishwashers (0.33 quad). Does not include water heating energy. 9) Commercial only; residential fan and pump energy use included proportionately in space heating and cooling. 10) Includes residential small electric devices, heating elements, motors, swimming pool heaters, hot tub heaters, outdoor grills, and natural gas outdoor lighting. Includes commercial service station equipment, ATMs, telecommunications equipment, medical equipment, pumps, emergency electric generators, combined heat and power in commercial buildings, and manufacturing performed in commercial buildings. 11) Energy adjustment EIA uses to relieve discrepancies between data sources. Energy attributable to the residential and commercial buildings sector, but not directly to specific end-uses.

Source(s): EIA, Annual Energy Outlook 2008, Mar. 2008, Tables A2, p. 117-119, Table A4, p. 122-

123, Table A5, p. 124-125, and Table A17, p. 143-144; and EIA, National Energy Modeling System for AEO 2008, Mar. 2008

2030 U.S. Buildings Energy End-Use Splits, by Fuel Type

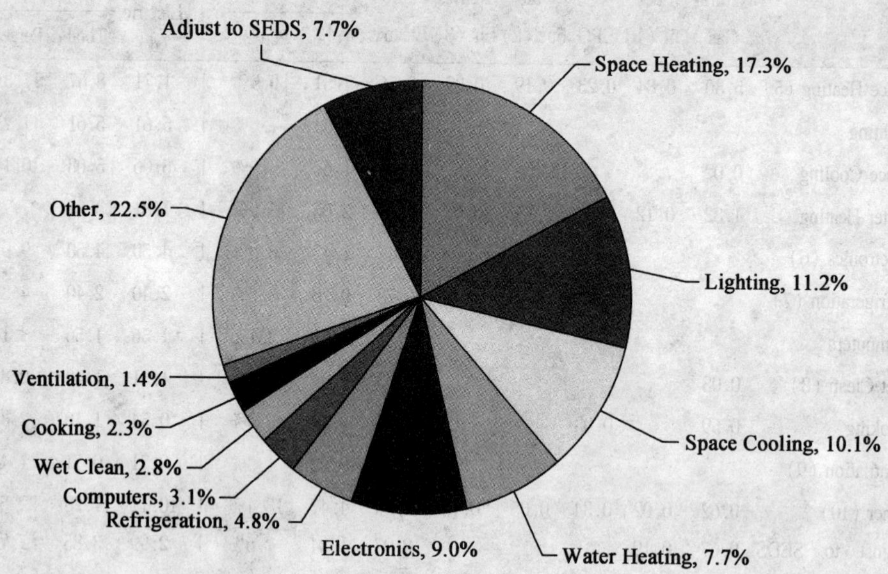

Appendix 2: Psychrometric Property Calculations

Environment Control for Animals and Plants*

(Albright, 1999. Not reproduced in its entirety)

2.1 Introduction

Predominantly, environmental control means control of the aerial environment. The fluid of interest is moist air, and the study of moist air is psychrometrics. Part of psychrometrics deals with methods to determine moist air properties at known states and part deals with determining properties at an unknown state after a defined change from a known state-the study of psychrometric processes.

Standard applications of gas laws can be found in basic physics and thermodynamics texts. Boyle's law, Charles' Law, Dalton's Law, and the perfect gas law apply with sufficient accuracy to atmospheric air to be acceptable for most environmental control applications. Errors introduced by assuming air acts as a perfect gas are generally less than 1%.

Dry air is defined as the state of air when all moisture and contaminants (pollutants, dust, etc.) have been removed; its composition is effectively constant. Carbon dioxide shows a seasonal and geographic variation and a long-term trend upward but today averages approximately 527 mg/kg. (Before the Industrial Revolution, the atmospheric carbon dioxide level is estimated to have been approximately 270 mg/L) The apparent molecular weight of standard dry air is 28.964 5 based on a weighted average of all components. Its gas constant is 287.055 J/(kg·K). Dry air is heavier than water vapor and lighter than carbon dioxide. The gas constant of water vapor is 461.52 J/(kg·K) and water's molecular weight is 18.015 34.

---

* Reprinted with permission from the American Society of Agricultural and Biological Engineers, 2950 Niles Road, St. Joseph, MI 49085-9659.

Standard conditions are frequently chosen for psychrometric calculations. The Standard Atmosphere is a typical choice; the Standard Atmosphere is at 15℃ temperature and 101.325 kPa pressure, which is sea level. Standard conditions for other elevations are in Appendix 2-1.

Appendix 2-1. Properties of the standard atmosphere at various elevations

Elevation above sea level (m)	Temperature, Celsius	Pressure (Pa)
-500	18.2	107 478
0	15.0	101 325
500	11.8	95 461
1 000	8.5	89 874
2 000	2.0	79 495
4 000	-11.0	61 640
5 000	-17.5	54 020

Adapted from the ASHRAE Handbook of Fundamentals, 1989.

2.2 Psychrometric Properties

Moist air is a mixture of dry air and water vapor. Numerous properties describe the state of air and its degree of saturation with water vapor. In most cases, knowledge of two of the properties permits determination of the others.

2.2.1 Dry-Bulb Temperature

The reference for specifying ordinary air temperature is dry-bulb temperature-the temperature sensed by an ordinary thermometer at thermal equilibrium with air.

2.2.2 Water Vapor Saturation Partial Pressure

The greatest amount of moisture which can be held by dry air occurs at saturation, defined as the state of equilibrium between moist air and free water on a flat surface. The air and water must be the same temperature-the temperature at which water would boil were it at a pressure equal to the water vapor saturation partial pressure. If water vapor partial pressure is less than the water vapor saturation partial pressure at the same temperature, the vapor is superheated.

Dry-bulb temperature determines the water vapor saturation partial pressure, P_{ws}

The water vapor saturation partial pressure can be determined (in Pa) from the following temperatures, T, in Kelvin:

$$\ln(P_{ws}) = A_1/T + A_2 + A_3T + A_4T^2 + A_5T^3 + A_6T^4 + A_7\ln(T) \quad (2\text{-}1)$$

In the dry-bulb temperature range from -100 to 0℃, water vapor saturation of air is over ice and the coefficients of Equation 2-1 have the following values:

$A_1 = -5.674\ 535\ 9 \times 10^3$

$A_2 = 6.392\ 524\ 7$

$A_3 = -9.677\ 843 \times 10^{-3}$

$A_4 = 0.622\ 157\ 0 \times 10^{-6}$

$A_5 = 2.074\ 782\ 5 \times 10^{-9}$

$A_6 = -0.948\ 402\ 4 \times 10^{-12}$

$A_7 = 4.163\ 501\ 9$

Over water, in the temperature range from 0 to 200℃, coefficients to calculate water vapor saturation partial pressure are:

$A_1 = -5.800\ 220\ 6 \times 10^3$

$A_2 = 1.391\ 499\ 3$

$A_3 = -48.640\ 239 \times 10^{-3}$

$A_4 = 41.764\ 768 \times 10^{-6}$

$A_5 = -14.452\ 093 \times 10^{-9}$

$A_6 = 0.0$

$A_7 = 6.545\ 967\ 3$

A graph of water vapor saturation partial pressure based on Equation 2-1 is in Figure 2-1.

2.2.3 Relative Humidity

The actual partial pressure of water vapor in moist but unsaturated air, P_w, is related to the property of air called the relative humidity, Φ, expressed as:

$$\Phi = P_w/P_{ws} \quad (2\text{-}2)$$

where the partial pressure of the water vapor and its partial pressure at saturation are defined at identical temperatures and atmospheric pressures. Relative humidity is a measure of the degree to which air is saturated, for Φ equals 0.0 for dry air and 1.0 when air is completely saturated.

Relative humidity also equals the mole fraction, x, of water vapor in a given air sample divided by the mole fraction were the air to be saturated at the same temperature and pressure,

$$\Phi = x_w / x_{ws} \tag{2-3}$$

2.2.4 Humidity Ratio

A second measure of the water vapor content of air is the humidity ratio, W, which is the mass of water vapor evaporated into a unit mass of dry air, kg/kg. The ratio of molecular masses of water vapor and (averaged) air is 18.015 34/28.964 5 =0.621 98. Thus,

$$W = 0.621\ 98 x_w / x_a \tag{2-4}$$

If air and water vapor are assumed to act as perfect gases,

$$P_a V = n_a RT, \tag{2-5}$$

$$P_w V = n_w RT, \text{ and} \tag{2-6}$$

$$PV = nRT, \tag{2-7}$$

where P is atmospheric pressure ($P = P_a + P_w$) and n is the number of moles of moist air ($n = n_a + n_w$).

Thus,

$$x_a = P_a / P;\ x_w = P_w / P; \tag{2-8}$$

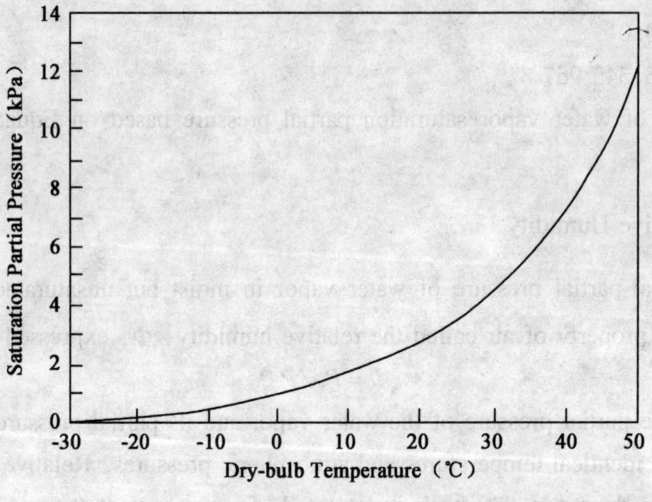

Figure 2-1. Water vapor saturation pressure, kPa, as a function of temperature, Celsius

and

$$W = 0.621\,98 P_w / (P - P_w) \tag{2-9}$$

$$W_s = 0.621\,98 P_{ws} / (P - P_{ws})$$

If dry-bulb temperature, atmospheric pressure, and relative humidity are known, the humidity ratio can be calculated. The equations can also be used to calculate relative humidity if dry-bulb temperature, atmospheric pressure, and humidity ratio are known. This is frequently useful in environmental control calculations, as shall be seen.

Example 2-1

Problem: Moist air at standard atmospheric pressure is at 20℃ dry-bulb temperature and 50% relative humidity. What is the humidity ratio of the air?

Solution: The humidity ratio is calculated from the partial pressure of the water vapor which can be obtained knowing the dry-bulb temperature and relative humidity.

From Equation 2-1 for a dry-bulb temperature of 293.15 K, the water vapor saturation partial pressure is

$$\begin{aligned} P_{ws} = & \exp(-5.800\,220\,6 \times 10^3/293.15 + 1.391\,499\,3) \\ & -(48.640\,239 \times 10^{-3})(293.15) + (41.764\,768 \times 10^{-6})(293.15)^2 \\ & -(14.452\,093 \times 10^{-9})(293.15)^3 + 6.545\,967\,3 \ln 293.15 \\ = & \exp(7.757) = 2\,339(\mathrm{Pa}) \end{aligned}$$

By the definition of relative humidity, Equation 2-2, the actual partial pressure of the water vapor is

$$P_w = \Phi P_{ws} = (0.50)(2\,339) = 1\,169 \text{ Pa} = 1.169 \text{ kPa} \tag{2-10}$$

The humidity ratio for an atmospheric pressure of 101.325 kPa is

$$W = 0.621\,98(1.169)/(101.325 - 1.169) = 0.007\,26 \text{ kg/kg}$$

2.2.5 Psychrometric Chart

Dry-bulb temperature and humidity ratio form the x and y axes of the psychrometric chart, a graph which can be used to determine many of an air sample's psychrometric properties if two of its properties are known. An outline of the psychrometric chart is in Figure 2-2.

Two standard psychrometric charts for sea level conditions are shown in Figure 2-3. (Charts for elevations above sea level are also available as engineering resources.) Dry-bulb temperature is the ordinate; humidity ratio is the abscissa.

These two scales are divided evenly which makes scales for other psychrometric properties nonlinear. Copies of the chart are available from ASHRAE, the American Society of Heating, Refrigerating, and Air Conditioning Engineers.

The line of water vapor saturation defines the upper boundary of the psychrometric chart, and lines of constant relative humidity parallel the saturation line (Note: The degree of saturation and water vapor partial pressure are not on the psychrometric chart and must be calculated if needed).

Although emphasis in this text will be on using equations and a computer for psychrometric calculations, it is necessary also to be able to use the psychrometric chart. Consider again Example 2-1, air at sea level contains a dry-bulb temperature of 20℃ and relative humidity of 50%. These conditions are found on Figure 2-3- normal temperatures. When the intersection of the two given conditions is located on the chart, and a horizontal line is followed to the abscissa, a humidity ratio of approximately 0.007 3 kg/kg is found.

Note that although dry-bulb temperature and humidity ratio form the x and y axes of the psychrometric chart, the independent variables used to graph the chart in the Mollier form are humidity ratio and enthalpy. One result of this is that lines of constant dry-bulb temperature are neither parallel to each other nor perpendicular to the x axis. Other forms of the psychrometric chart can be found which are graphed using other independent variables, but the Mollier form is the ASHRAE standard. Fewer thermodynamic approximations are required when the chart is drawn in this form.

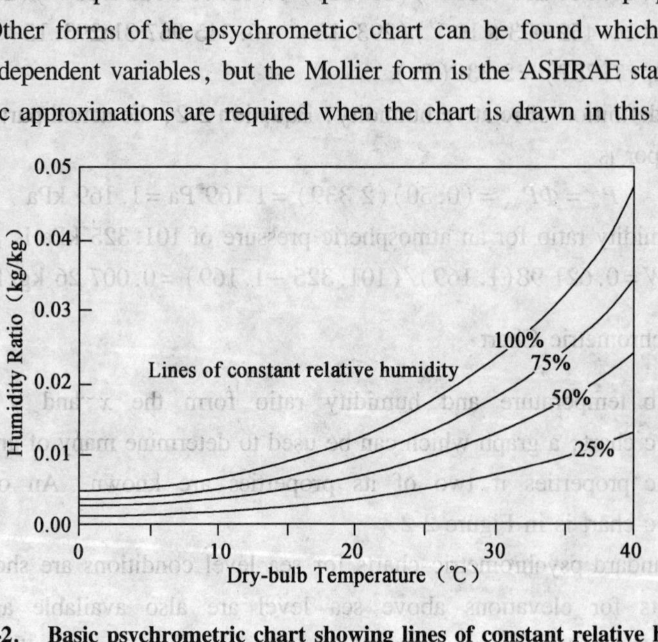

Figure 2-2. Basic psychrometric chart showing lines of constant relative humidity

2.2.6 Degree of Saturation

Relative humidity was defined as the ratio of partial vapor pressures and is a measure of the extent to which air is saturated with water vapor. The degree of saturation, μ, is a separate means to describe the extent of saturation. It is defined as

$$\mu = W/W_s \tag{2-11}$$

which is not the same as relative humidity. The two are related by

$$\Phi = \mu/[1-(1-\mu)(P_{ws}/P)], \text{and} \tag{2-12a}$$

$$\mu = \Phi(1-P_{ws}/P)/(1-\Phi P_{ws}/P) \tag{2-12b}$$

The degree of saturation and relative humidity are equivalent for dry air and saturated air, but they differ at intermediate moisture contents because of the nonlinear relationship between humidity ratio and water vapor partial pressure.

Example 2-2

Problem: To continue the example of air at 20℃ and 50% relative humidity, determine the degree of saturation of the air.

Solution: The water vapor saturation partial pressure was calculated at 2.339 kPa and the actual water vapor partial pressure as 1.169 kPa. For the Standard Atmosphere, W was previously calculated as 0.007 26 kg/kg. Equation 2-9 can be used again to find

$$W_s = 0.621\,98(2.339)/(101.325-2.339) = 0.014\,70 \text{ kg/kg}$$

and by Equation 2-11, the degree of saturation, 0.494, is slightly different from the relative humidity. To complete the circle, Equation 2-12a yields a relative humidity value of 0.5.

2.2.7 Specific Volume and Density

A physical property of moist air useful for environmental calculations is density, ρ, or its inverse, specific volume, v. Specific volume is based on a unit mass of dry air, thus,

$$v = V_a/M_a = V/28.964\,5\, n_a \tag{2-13}$$

The perfect gas relationship, with $p = p_a + p_w$, leads to

$$v = R_a T/(P-P_w) \tag{2-14}$$

which, when combined with the definition of the humidity ratio, provides the following computational equation for specific volume of dry air, and through its inverse, density:

$$v_{\text{dry air}} = (1/P)R_aT(1+1.6078W) \tag{2-15}$$

Note: In psychrometric equations, specific quantities are based on a unit mass of dry air, so, although units of ν are in m^3/kg dry air, this is volume of moist air per unit mass of dry air. Dry air forms the basis of specific quantities because it is unchanged as the humidity changes during a psychrometric process. The value of specific volume used for environmental control calculations is typically based on moist air. Correcting the specific volume of dry air to the specific volume of moist air leads to

$$v = [(1/P)R_aT(1+1.6078W)]/(1+W) \tag{2-16}$$

The psychrometric chart includes specific volume of moist air as a parameter. Steep, diagonal lines from upper left to lower right define specific volume. Air density, ρ, is its reciprocal. Note: In Equation 2-15 and Equation 2-16, atmospheric pressure is in Pascals not kilopascals.

Example 2-3

Problem: For the previous example of 20℃ and 50% relative humidity, the psychrometric chart shows a specific volume of approximately 0.84 m^3/kg. Determine the specific volume of the moist air using psychrometric equations instead.

Solution: Applying Equation 2-16 yields the following estimate of specific volume of the moist air:

$$v = \frac{(1/101\,325)\times(287.055)\times(293.15)\times[1+(1.6078)\times(0.00726)]}{1+0.00726}$$

$$=0.83\ (m^3/kg)$$

2.2.8 Dew Point Temperature

When moist air is cooled sufficiently, water vapor condenses. Rain, snow, fog, and dew are formed this way. The temperature at which condensation occurs is the dew point temperature, T_d. When the state of air is below the saturation line on the psychrometric chart (as it is in normal situations), the dew point temperature is less than the dry-bulb temperature. That is, air in typical situations must be cooled at least a little before water begins to condense. The difference between dry-bulb and dew point temperatures is determined by the moisture content of the initial air sample relative to the moisture content at saturation at the same dry bulb temperature. Of course, at saturation the dew point and dry bulb temperatures are the same.

When air is cooled, initially, there is no moisture content change. The relative humidity rises because cooler air has less moisture-holding capacity than does warmer air. However, the humidity ratio remains unchanged until air is cooled to the point of saturation, where water begins to condense out of the air if cooling is below the dew point temperature. If this reasoning is applied to the psychrometric chart, the dew point temperature is at the intersection of the saturation line and the line of constant humidity ratio (a horizontal line) which passes through the state defining the original condition of the air.

Because dew point temperature is a function of the partial pressure of water vapor in the air, it may be calculated from one of the following equations.

For conditions of frost between −60℃ and 0℃, the dew point in *Celsius* is

$$T_d = -60.45 + 7.0322 \ln(P_w) + 0.3700(\ln(P_w))^2. \qquad (2\text{-}17)$$

For temperatures between 0 and 70℃

$$T_d = -35.957 - 1.8726 \ln(P_w) + 1.1689 (\ln(P_w))^2, \qquad (2\text{-}18)$$

where P_w has units of Pascals.

***Example** 2-4*

Problem: Continuing with the example of air at 20℃ and 50% relative humidity, the psychrometric chart shows the dew point temperature to be approximately 9℃. Check this value by calculation.

Solution: For a partial pressure of water vapor of 1 169 Pa, as calculated in Example 2-1, the dew point can be found using Equation 2-18.

$$T_d = -35.957 - 1.8726 \ln(1169) + 1.1689(\ln(1169))^2 = 9.14(℃).$$

2.2.9 Enthalpy

The enthalpy, h, of moist air is a property useful in quantifying psychrometric processes which involve thermal energy exchanges. Enthalpy is an extensive property, thus, the enthalpy of a mixture equals the sum of enthalpies of the parts - the dry air, h_a, and water vapor, h_w:

$$h = h_a + h_w \qquad (2\text{-}19)$$

Enthalpy in psychrometrics is referenced to 0℃ (in the SI system) and water is assumed to be liquid at that temperature. The specific heat of dry air is 1.006 kJ/(kg·K) and the specific heat of water vapor is 1.805 kJ/(kg·K). The specific heat of dry air actually varies from 0℃ to 1.009 at 50℃, but the change is so slight it usually is ignored.

Note: Psychrometric calculations in the IP system of units reference enthalpy to 0℉. Conversion of enthalpy values from one set of units to the other is not straightforward. When converting from one system of units to the other, the enthalpy difference between 0℃ and 0℉ must always be included in calculations of enthalpy. Calculations of enthalpy changes do not require this correction.

The heat of vaporization of water at 0℃ is 2 501 kJ/kg. Heat of vaporization, kJ/kg, is a function of temperature and can be described by

$$h_{fg} = 2\,501 - 2.42T \tag{2-20}$$

for dry-bulb temperatures, T, between 0℃ and 65℃. Because of its weak dependence on temperature, the heat of vaporization is often treated as a constant, 2 501 kJ/kg.

Using the data as given, enthalpy of moist air in kJ/kg of dry air can be calculated if dry-bulb temperature and the humidity ratio are known.

$$h = 1.006T + W(2\,501 + 1.805T) \tag{2-21}$$

It should be noted that in Equation 2-21, the combination $1.006 + 1.805W$ is frequently treated as the specific heat [kJ/(kg · K)] of moist air. However, in many practical applications, the value 1.006 kJ/(kg · K) is used for convenience to represent the specific heat of moist air. The error in this assumption is small. Note that in Equation 2-21, enthalpy is calculated assuming the water contained in the air evaporates at 0℃ and is then heated, along with the air, to T. This is appropriate as enthalpy is a function of state, not path.

In Psychrometric Charts lines of constant enthalpy are included and are the solid lines from the lower right to the upper left corner of the chart.

Example 2-5

Problem: Returning to the example of 20℃ and 50% relative humidity, the Psychrometric chart can be used to estimate an enthalpy content relative to 0℃ of 39 kJ/kg of dry air. Check this value by calculation.

Solution: Equation 2-21 can be used to obtain a computed value:

$$h = 1.006(20) + 0.007\,26(2\,501 + (1.805) \times (20) = 38.5 \text{ kJ/kg}$$

2.2.10 Wet-Bulb Temperature

Another measure of the moisture content of air is the thermodynamic wet-bulb temperature. For any sample of moist air, a mental experiment could be performed which would isolate the air adiabatically from its surroundings, but permit contact with a water source on a flat surface (to eliminate surface tension effects) for water vapor exchange, with the limitation that thermal energy needed to evaporate water would come only from the air. Except for energy used for evaporation, no other thermal energy could be exchanged between the air and water.

If the air sample were not initially saturated, evaporation would occur with thermal energy from the air being the source of energy for evaporation. Water vapor would mix with the air and the humidity ratio would increase. The process would continue until the air became saturated. The air would have cooled, for sensible heat would have been transformed into latent heat. The cooling process would proceed until saturation. If the air were very dry initially, a great deal of cooling would occur. If the air were near saturation, little cooling would be possible. The limiting temperature, at which the air would become saturated, is the wet-bulb temperature.

The assumption of adiabatic conditions infers enthalpy is conserved in the evaporation process. Enthalpy (in the form of sensible heat) from the air used for evaporation returns to the air as latent heat. However, although enthalpy is conserved overall, the air gains enthalpy slightly in the process, and the water loses a like amount. The difference is the enthalpy content of the evaporated water prior to evaporation.

The wet-bulb temperature for given conditions of moist air can be determined from the psychrometric chart. Diagonal dashed lines from the lower right to the upper left on the chart intersect the saturation line at the wet-bulb temperature corresponding to the initial conditions. There is a unique wet-bulb temperature for any given set of initial conditions although the reverse is not true. A given wet-bulb temperature applied to any of the combinations of conditions which lie along the wet-bulb line. Note that lines of constant wet-bulb temperature do not correspond to lines of constant enthalpy. The difference is the enthalpy content of water prior to evaporation. However, the two lines are nearly parallel, and on some psychrometric charts they are used interchangeably (sometimes with a correction factor introduced).

Wet-bulb temperature can be determined using equations, but the computational process is iterative. The process of adiabatic saturation of air is described by

$$h_s^* = h + h_w^* (W_s^* - W) \qquad (2\text{-}22)$$

which is a statement of the enthalpy change of air as it saturates adiabatically. The humidity ratio starts at W and in the constant pressure process increases from W to W_s^* at saturation (the $*$ superscript indicates conditions at the wet-bulb temperature) while enthalpy of the air increases slightly from h to h_s^*. Saturation is at the wet-bulb temperature, T^* in ℃.

Enthalpy of water, h_w^*, is found from

$$h_w = 4.186T; \text{ and } h_w^* = 4.186T^* \qquad (2\text{-}23)$$

A procedure to determine wet-bulb temperature is to search for a temperature, T^*, such that Equation 2-22 is satisfied. An ad hoc search procedure is to begin the search slightly above the air's dry-bulb temperature, and decrement the candidate temperature until the equation is satisfied to within an acceptable level of precision. Values for atmospheric pressure, initial dry-bulb temperature, initial enthalpy, and initial humidity ratio are required to begin the process. Accelerated search techniques sweeping in alternate directions with ever-decreasing step sizes, or using the bisection method, can reduce the time required for convergence, but searching in one direction with small decrements of temperature will lead to a solution as long as the initial direction of search is in the proper direction.

Appendix 3: Appendices from Louis D. Albright's Book

Environmental Control of Animals and Plants, 1990 ed. Reprinted with permission from the Association of American Society of Agricultural and Biological Engineers (ASABE), 2950 Niles Road, St. Joseph, MI 49085-9659. Library of Congress Catalog Card Number 90-062260.

Appendix 3-1. Selected Values of Thermal Conductivity[a]

Material	Thermal Conductivity (W/m·K)
Metals	
Aluminum (alloy 1100)	221
Brass, red (85% Cu, 15% Zn)	150
Brass, yellow (65% Cu, 35% Zn)	120
Copper (electrolytic)	393
Gold	297
Iron, cast	47.7 (327 K)
Iron, wrought	60.4
Lead	34.8
Nickel	59.5
Silver	424
Steel, mild	45.3
Tin	64.9
Zinc, galvanizing	110

Continued

Material	Thermal Conductivity (W/m·K)
Wood	
Ash, white	0.172
Elm, American	0.153
Fir, white	0.12
Mahogony	0.13
Maple, sugar	0.187
Oak, white	0.176
Pine, white	0.11
Spruce	0.11
Other	
Brick, building	0.7
Cardboard	0.07
Cellulose	0.057
Charcoal, wood	0.05 (473 K)
Concrete stone	0.93
Cork granulated	0.048 (268 K)
Cotton, fiber	0.042
Earth, dry and packed	0.064
Glass, soda-lime	1.0 (366 K)
Ice, 0℃	2.24
Leather	0.16
Marble	2.6
Paper	0.13
Plaster	0.74 (348 K)
Sand, dry	0.33
Sawdust	0.05
Snow, fresh at 32℃	0.598

Selected from the 1985 ASHRAE Handbook of Fundamentals. a. values at room temperature unless noted in parentheses.

Appendix 3-2 Design Heat Transmission Coefficients Taken from the 1985 ASHRAE Handbook of Fundamentals, American Society of Heating, Refrigerating and Air Conditioning Engineers, Atlanta GA. (Used by permission)

Thermal Properties of Typical Building and Insulating Materials-Design Values[a]

Description	Density (kg/m³)	Conductivity (λ)(W/m · ℃)	Conductance (C)(W/m² · ℃)	Resistance (R) Per meter thickness (1/λ)(m · ℃/W)	Resistance (R) For thickness listed(1/C) (m² · ℃/W)	Specific Heat (kJ/kg · ℃)
Building Board						
Boards, Panels, Subflooring, Sheathing						
Woodboard Panel Products						
Asbestos-cement board	1920	0.576	—	1.74	—	1.01
Asbestos-cement board ······ 3.18 mm	1920	—	187.4	—	0.005	
Asbestos-cement board ······ 6.35 mm	1920	—	93.72	—	0.011	
Gypsum or plaster board ······ 9.53 mm	800	—	17.61	—	0.056	1.09
Gypsum or plaster board ······ 12.70 mm	800	—	12.61	—	0.079	
Gypsum or plaster board ······ 15.88 mm	800	—	10.11	—	0.099	
Plywood (Douglas Fir)	544	0.115	—	8.68	—	1.22
Plywood (Douglas Fir) ······ 6.35 mm	544	—	18.18	—	0.055	
Plywood (Douglas Fir) ······ 9.53 mm	544	—	12.10	—	0.083	
Plywood (Douglas Fir) ······ 12.70 mm	544	—	9.09	—	0.11	
Plywood (Douglas Fir) ······ 15.88 mm	544	—	7.33	—	0.14	
Plywood or wood panels ······ 19.05 mm	544	—	6.08	—	0.16	1.22
Vegetable Fiber Board						
Sheathing, regular density ······ 12.70 mm	288	—	4.32	—	0.23	1.30

Continued

Description	Density (kg/m^3)	Conductivity (λ)(W/m · ℃)	Conductance (C)(W/m^2 · ℃)	Resistance (R) Per meter thickness ($1/\lambda$)(m · ℃/W)	Resistance (R) For thickness listed($1/C$)(m^2 · ℃/W)	Specific Heat (kJ/kg · ℃)
················ 19.84 mm	288	—	2.78	—	0.36	
Sheathing intermediate density ········· 12.70 mm	352	—	4.66	—	0.21	1.30
Nail-base sheathing ························ 12.70 mm	400	—	5.00	—	0.20	1.30
Shingle backer ······························ 9.53 mm	288	—	6.02	—	0.17	1.30
Shingle backer ······························ 7.94 mm	288	—	7.27	—	0.14	
Sound deadening board ··················· 12.70 mm	240	—	4.20	—	0.24	1.26
Tile and lay-in panel, plain or acoustic ············	288	0.058	—	17.35	—	0.59
··· 12.70 mm	288	—	4.54	—	0.22	
··· 19.05 mm	288	—	3.01	—	0.33	
Laminated paperboard ································	480	0.072	—	13.88	—	1.38
Homogeneous board from repulped paper ··········	480	0.072	—	13.88	—	1.17
Hardboard						
Medium density ··	800	0.105	—	9.51	—	1.30
High density, service temp. service underlay ·····	880	0.118	—	8.47	—	1.34
High density, std. tempered ··························	1008	0.144	—	6.94	—	1.34
Particleboard						
Low density ··	592	0.078	—	12.84	—	1.30
Medium density ··	800	0.135	—	7.36	—	1.30
High density ···	1000	0.170	—	5.90	—	1.30

Continued

Description	Density (kg/m³)	Conductivity (λ)(W/m · ℃)	Conductance (C)(W/m² · ℃)	Resistance (R) Per meter thickness (1/λ)(m · ℃/W)	Resistance (R) For thickness listed(1/C) (m² · ℃/W)	Specific Heat (kJ/kg · ℃)
Underlayment ………… 15.88 mm	640	—	6.93	—	0.14	1.22
Wood subfloor ………… 19.05 mm		—	6.02	—	0.17	1.38
Building Membrane						
Vapor-permeable felt …………	—	—	94.86	—	0.011	
Vapor-seal, 2 layers of mopped 0.73 kg/m² felt …………	—	—	47.43	—	0.021	
Vapor-seal, plastic film …………	—	—	—	—	Negl.	
Finish Flooring Materials						
Carpet and fibrous pad …………	—	—	2.73	—	0.37	1.42
Carpet and rubber pad …………	—	—	4.60	—	0.22	1.38
Cork tile ………… 3.18 mm	—	—	20.45	—	0.049	2.01
Terrazzo ………… 25.40 mm	—	—	71.00	—	0.014	0.80
Tile-asphalt, Linoleum, vinyl, rubber …………	—	—	113.6	—	0.009	1.26
vinyl asbestos …………						1.01
ceramic …………						0.80
Wood, hardwood finish ………… 19.05 mm			8.35		0.12	
Insulating Materials						
Blanket and Batt						
Mineral Fiber, fibrous form processed from rock, slag, or glass						
approx.[c] 76.2 – 101.6 mm …………	4.8 – 32.0	—	0.52	—	1.94[b]	
approx.[c] 88.9 mm …………	4.8 – 32.0	—	0.44	—	2.29[b]	
approx.[c] 139.7 – 165.1 mm …………	4.8 – 32.0	—	0.30	—	3.34[b]	

Continued

Description	Density (kg/m³)	Conductivity (λ)(W/m · ℃)	Conductance (C)(W/m² · ℃)	Resistance (R) Per meter thickness (1/λ)(m · ℃/W)	Resistance (R) For thickness listed(1/C) (m² · ℃/W)	Specific Heat (kJ/kg · ℃)
approx. [c] 152.4 – 177.8 mm	4.8 – 32.0	—	0.26		3.87[b]	
approx. [c] 215.9 – 228.6 mm	4.8 – 32.0		0.19		5.28[b]	
approx. 304.8 mm	4.8 – 32.0	—	0.15		6.69[b]	
Board and Slabs						
Cellular glass	136	0.050	—	19.85	—	0.75
Glass fiber, organic bonded	64 – 144	0.036	—	27.76	—	0.96
Expanded perlite, organic bonded	16.0	0.052	—	19.29	—	1.26
Expanded rubber (rigid)	72.0	0.032	—	31.58	—	1.68
Expanded polystyrene extruded						
Cut cell surface	28.8	0.036	—	27.76	—	1.22
Smooth skin surface	28.8 – 56.0	0.029	—	34.70	—	1.22
Expanded polystyrene, molded beads	16.0	0.037	—	23.25	—	—
	20.0	0.036	—	27.76	—	—
	24.0	0.035	—	28.94	—	—
	28.0	0.035	—	28.94	—	—
	32.0	0.033	—	30.19	—	—
Cellular polyurethane[c] (R-11 exp.) (unfaced)	24.0	0.023	—	43.38	—	1.59
Foil-faced, glass fiber-reinforced cellular						
Polyisocyanurate (R-11 exp.)[d]	32.0	0.020	—	49.97	—	0.92

Continued

Description	Density (kg/m³)	Conductivity (λ)(W/m · ℃)	Conductance (C)(W/m² · ℃)	Resistance (R) Per meter thickness (1/λ)(m · ℃/W)	For thickness listed(1/C) (m² · ℃/W)	Specific Heat (kJ/kg · ℃)
Nominal 12.70 mm		—	1.58	—	0.63	
Nominal 25.40 mm		—	0.79	—	1.27	
Nominal 50.80 mm		—	0.39	—	2.53	
Mineral fiber with resin binder	240	0.042	—	23.94	—	0.71
Mineral fiberboard, wet felted						
Core or roof insulation	256 – 272	0.049	—	20.40	—	
Acoustical tile	288	0.050	—	19.85	—	0.80
Acoustical tile	336	0.053	—	18.74	—	
Mineral fiberboard, wet molded						
Acoustical tile[c]	368	0.060	—	16.52	—	0.59
Wood or cane fiberboard						
Acoustical tile[c] 12.70 mm	—	—	4.54	—	0.22	1.30
Acoustical tile[c] 19.05 mm	—	—	3.01	—	0.33	
Interior finish (plank, tile)	240	0.050	—	19.85	—	1.34
Cement fiber slabs (shredded wood with Portland cement binder)	400 – 432	0.072 – 0.070	—	13.88 – 13.12	—	—
Cement fiber slabs (shredded wood with magnesia oxysulfide binder)	352	0.082	—	12.15	—	1.30
Loose fill						
Cellulosic insulation (milled paper or wood pulp)	36.8 – 51.2	0.039 – 0.046	—	25.68 – 21.72	—	1.38

Continued

Description	Density (kg/m^3)	Conductivity (λ)(W/m · ℃)	Conductance (C)(W/m^2 · ℃)	Resistance (R) Per meter thickness ($1/\lambda$)(m · ℃/W)	Resistance (R) For thickness listed($1/C$) (m^2 · ℃/W)	Specific Heat (kJ/kg · ℃)
Sawdust or shavings	128 – 240	0.065	—	15.41	—	1.38
Wood fiber, softwoods	32.0 – 56.0	0.043	—	23.11	—	1.38
Perlite, expanded						
	32.0 – 65.6	0.039 – 0.045		25.68 – 22.90		
	65 – 118	0.045 – 0.052		22.90 – 19.43		
	118 – 176	0.052 – 0.060		19.43 – 16.66		
Mineral fiber (rock, slag or glass)						
approx. 95.3 – 127.0 mm	9.6 – 32.0	—	—		1.94	0.71
approx. 165.1 – 222.3 mm	9.6 – 32.0	—	—		3.34	
approx. 190.5 – 254.0 mm	9.6 – 32.0	—	—		3.87	
approx. 260.4 – 349.3 mm	9.6 – 32.0	—	—		5.28	
Mineral fiber (rock, slag or glass)						
approx. 83.8 mm (closed sidewall application)	32.0 – 56.0	—	—	—	2.46	
Vermiculite, exfoliated	112 – 131	0.068	—	14.78	—	1.34
	64.0 – 96.0	0.063	—	15.75	—	
Field applied						
Polyurethane foam	24.0 – 40.0	0.023 – 0.026	—	43.38 – 36.50	—	
Ureaformaldehyde foam	11.2 – 25.6	0.032 – 0.040	—	24.78 – 31.58	—	

Continued

Description	Density (kg/m³)	Conductivity (λ) (W/m · ℃)	Conductance (C) (W/m² · ℃)	Resistance (R) Per meter thickness (1/λ) (m · ℃/W)	For thickness listed (1/C) (m² · ℃/W)	Specific Heat (kJ/kg · ℃)
Spray cellulosic fiber base	32.0 –96.0	0.035 –0.043	—	13.11 –28.94	—	
Plastering Materials						
Cement plaster, sand aggregate	1865	0.720	—	1.39	—	0.84
Sand aggregate .... 9.53 mm	—	—	75.54	—	0.014	0.84
Sand aggregate .... 19.05 mm	—	—	37.83	—	0.026	0.84
Gypsum plaster:						
Lightweigh aggregate .... 12.70 mm	720	—	17.72	—	0.056	
Lightweigh aggregate .... 15.88 mm	720	—	15.17	—	0.069	
Lightweight agg. on metal lath .... 19.05 mmin.	—	—	12.10	—	0.083	
Perlite aggregate	720	0.216	—	4.65	—	1.34
Plastering Materials						
Sand aggregate	1 680	0.806	—	1.25	—	0.84
Sand aggregate .... 12.70 mm	1 680	—	63.05	—	0.016	
Sand aggregate .... 15.88 mm	1 680	—	51.69	—	0.019	
Sand aggregate on metal lath .... 19.05 mm	—	—	43.74	—	0.023	
Vermiculite aggregate	720	0.245	—	4.09	—	
Masonry Materials						
Concretes						
Cement mortar	1856	0.720	—	1.39	—	

Continued

Description	Density (kg/m³)	Conductivity (λ)(W/m · ℃)	Conductance (C)(W/m² · ℃)	Resistance (R) Per meter thickness (1/λ)(m · ℃/W)	Resistance (R) For thickness listed(1/C) (m² · ℃/W)	Specific Heat (kJ/kg · ℃)
Gypsum-fiber concrete 87.5% gypsum, 12.5% wood chips	816	0.239	—	4.16	—	0.88
Lightweight aggregates including	1920	0.749	—	1.32	—	
Expanded shale, clay or slate; expanded	1600	0.518	—	1.94	—	
slags; cineders; pumice; vermiculite;	1280	0.360	—	2.78	—	
also cellular concretes	960	0.245	—	4.09	—	
	640	0.166	—	5.97	—	
	480	0.130	—	7.70	—	
	320	0.101		9.92		
perlite, expanded	640	0.134		7.50		
	480	0.102		9.79		
	320	0.072		13.88		1.34
Sand and gravel or stone aggregate (oven dried)	2240	1.296	—	0.76		0.92
Sand and graved or stone aggregate (not dried)	2240	1.728	—	0.56		
Stucco	1856	0.720	—	1.39		
Masonry Units						
Brick, common[f]	1920	0.720	—	1.39	—	0.80

Continued

Description	Density (kg/m³)	Conductivity (λ)(W/m · ℃)	Conductance (C)(W/m² · ℃)	Resistance (R) Per meter thickness (1/λ)(m · ℃/W)	Resistance (R) For thickness listed(1/C) (m² · ℃/W)	Specific Heat (kJ/kg · ℃)
Brick, face[f] …………	2080	1.296	—	0.76	—	
Clay tile, hollow:						
1 cell deep ………… 76.2 mm	—	—	7.10	—	0.14	0.88
1 cell deep ………… 101.6 mm	—	—	5.11	—	0.20	
2 cells deep ………… 152.4 mm	—	—	3.75	—	0.27	
2 cells deep ………… 203.2 mm	—	—	3.07	—	0.33	
2 cells deep ………… 254.0 mm	—	—	2.56	—	0.39	
3 cells deep ………… 304.8 mm	—	—	2.27	—	0.44	
Concrete blocks, three oval core:						
Sand and gravel aggregate ………… 101.6 mm	—	—	7.95	—	0.12	0.92
………… 203.2 mm	—	—	5.11	—	0.20	
………… 304.8 mm	—	—	4.43	—	0.23	
Cinder aggregate ………… 76.2 mm	—	—	6.59	—	0.15	0.88
………… 101.6 mm	—	—	5.11	—	0.20	
………… 203.2 mm	—	—	3.29	—	0.30	
………… 304.8 mm	—	—	3.01	—	0.33	
Lightweight aggregate ………… 76.2 mm	—	—	4.49	—	0.22	0.88
expanded shale, clay, slate ………… 101.6 mm	—	—	3.81	—	0.26	
or slag; pumice ………… 203.2 mm	—	—	2.84	—	0.35	

Continued

Description	Density (kg/m³)	Conductivity (λ)(W/m·℃)	Conductance (C)(W/m²·℃)	Resistance (R) Per meter thickness (1/λ)(m·℃/W)	Resistance (R) For thickness listed (1/C) (m²·℃/W)	Specific Heat (kJ/kg·℃)
·························· 304.8 mm	—	—	2.50	—	0.40	
Concrete blocks, rectangular core[g]						
Sand and gravel aggregate						
2 core,[h] ······················ 203.2 mm, 16.3 kg	—	—	5.54	—	0.18	0.92
Same with filled cores[l] ································	—	—	2.95	—	0.34	0.92
Lightweight aggregate (expanded shale, clay, slate or slag, pumice):						
3 core,[h] ······················ 152.4 mm, 8.6 kg	—	—	3.46	—	0.29	0.88
Same with filled cores[i] ····························	—	—	1.87	—	0.53	
2 core,[h] ···················· 203.2 mm, 10.9 kg	—	—	2.61	—	0.38	
Same with filled cores[i] ····························	—	—	1.14	—	0.89	
3 core,[h] ···················· 304.8 mm, 17.3 kg	—	—	2.27	—	0.44	
Same with filled cores[i] ····························	—	—	0.97	—	1.02	
Stone, lime or sand ····································	—	1.800	—	0.56	—	0.80
Gypsum partition tile:						
76.2 · 304.8 · 762.0 mm, solid ······················	—	—	4.49	—	0.22	0.80
76.2 · 304.8 · 762.0 mm, 4 – cell ····················	—	—	4.20	—	0.24	
101.6 · 304.8 · 762.0 mm, 3 – cell ················	—	—	3.41	—	0.29	

Continued

Description	Density (kg/m^3)	Conductivity (λ)($W/m \cdot ℃$)	Conductance (C)($W/m^2 \cdot ℃$)	Resistance (R)		Specific Heat ($kJ/kg \cdot ℃$)
				Per meter thickness ($1/\lambda$)($m \cdot ℃/W$)	For thickness listed($1/C$) ($m^2 \cdot ℃/W$)	
Metals						
(See Chapter 39, Table 3)						
Roofing						
Asbestos-cement shingles	1920	—	27.04	—	0.037	1.01
Asphalt roll roofing	1120	—	36.92	—	0.026	1.51
Asphalt shingles	1120	—	12.89	—	0.077	1.26
Built-up ... 9.53 mm	1120	—	17.04	—	0.058	1.47
Slate ... 12.70 mm	—	—	113.6	—	0.009	1.26
Wood shingles, plain and plastic film faced	—	—	6.02	—	0.17	1.30
Siding Materials (on flat surface)						
Shingles						
Asbestos-cement	1920	—	26.98	—	0.037	
Wood, 406.4 mm. 190.5 mm exposure	—	—	6.53	—	0.15	1.30
Wood, double, 406.4 mm, 304.8 mm exposure	—	—	4.77	—	0.21	1.17
Wood, plus insul. backer board, 7.94 mm	—	—	4.03	—	0.25	1.30
Siding						
Asbestos-cement, 6.35 mm, lapped	—	—	27.04	—	0.037	1.01
Asphalt roll siding	—	—	36.92	—	0.026	1.47
Asphalt insulating siding (12.70 mm bed.)	—	—	3.92	—	0.26	1.47
Hardboard asiding, 11.11 mm	640	0.215	—	4.65		1.17
Wood, brop, 25.4 · 203.2 mm	—	—	7.21	—	0.14	1.17
Wood, bevel, 12.7 · 203.2 mm, lapped	—	—	6.99	—	0.14	1.17

Continued

Description	Density (kg/m^3)	Conductivity $(\lambda)(W/m \cdot ℃)$	Conductance $(C)(W/m^2 \cdot ℃)$	Resistance (R) Per meter thickness $(1/\lambda)(m \cdot ℃/W)$	Resistance (R) For thickness listed$(1/C)$ $(m^2 \cdot ℃/W)$	Specific Heat $(kJ/kg \cdot ℃)$
Wood, bevel, 19.1 · 254.0 mm, lapped	—	—	5.40	—	0.18	1.17
Wood, plywood, 9.53 mm, lapped	—	—	9.03	—	0.10	1.22
Aluminum or Steel[i], over sheathing						
Hollow-backed	—	—	9.14	—	0.11	1.22
Insulating-board backed nominal 9.53 mm	—	—	3.12	—	0.32	1.34
Insulating-board backed nominal 9.53 mm, foil backed			1.93		0.52	
Architectural glass	—	—	56.80	—	0.018	0.84
Woods (12% Moisture Coutent)[k,J]						
Hardwoods						1.63
Oak	659 – 749	0.169 – 0.180	—	6.18 – 5.55	—	
Birch	682 – 726	0.167 – 0.176	—	6.04 – 5.69	—	
Maple	637 – 704	0.157 – 0.171	—	6.52 – 6.11	—	
Ash	614 – 670	0.153 – 0.164	—	6.52 – 6.11	—	
Softwoods						1.63
Southern Pine	570 – 659	0.144 – 0.161	—	6.94 – 6.18	—	
Douglas Fir-Larch	536 – 581	0.137 – 0.145	—	7.36 – 6.87	—	
Southern Cypress	502 – 514	0.130 – 0.132	—	7.70 – 7.56	—	
Hem-Fir, Spruce-Pine-Fir	392 – 502	0.107 – 0.130	—	9.37 – 7.70	—	
West Coast Woods, Cedars	347 – 502	0.098 – 0.130	—	10.27 – 7.70	—	
California Redwood	392 – 448	0.107 – 0.118	—	9.37 – 8.47	—	

Notes

[a] Except where otherwise noted, all values are for a mean temperature of 23.9℃. Representative values for dry materials, selected by ASHRAE TC 4.4 are intended as design (not specification) values for materials in normal use. Insulation materials in actual service may have thermal values that vary from design values depending on their in-situ properties (e.g., density and moisture content). For properties of a particular product, use the value supplied by the manufacturer or by usbiased tests.

[b] Does not include paper backing and facing, if any. Where insulation forms a boundary (reflective or otherwise) of an air space, see Tables 1, 2A and 2B for the insulating value of an air space with the appropriate effective emittance and temperature conditions of the space.

[c] Values ar for aged, unfaced, board stock. For change in conductivity with age of expanded urethane, see Chapter 20, Factors Affecting Thermal Conductivity.

[d] Time-aged values for board stock with gas-barrier quality (0.025 mm thickness or greater) aluminum foil facers on two major surfaces.

[e] Insulating values of acoustical tile vary, depending on density of the board and on type, size and depth of perforations.

[f] Face brick and common brick do not always have these specific densities, When density differs from that shown, there is a change in thermal conductivity.

[g] Data on rectangular core concrete blocks differ from the above data on oval core blocks, due to core configuration, different mean temperatures, and possibly differenees in unit weights. Weight data on the oval core blocks tested are not available.

[h] Weights of units approximately 193.7 mm high and 400.1 mm long. These weights are given as a means of describing the blocks tested, but conductance values are all for 0.093 m^2 of area.

[i] Vermiculite, perlite, or mineral wool insulation. Where insulation is used, vapor barriers or other precautions must be considered to keep insulation dry.

[j] Values for metal siding applied over flat surfaces vary widely, depending on amount of ventilation of air space beneath the siding; whether air space is reflective or nonreflective; and on thickness, type, and application of insulating backing-board used. Values given are averages for use as design guides, and were obtained from several guarded hotbox lests (ASTM C236) or calibrated hotbox (BSS 77) on hollow-backed types and types made using backing-boards of wood fiber, foamed plasitic, and glass fiber. Departures of ±50% or more from the values given may occur.

[k] Forest Products Laboratory Wood Handbook, U.S. Dept, of Agriculture 72, 1974, Tables 3 and 4.

[l] L. Adams: Supporting cryogenic equipment with wood (*Chemical Engineering*, May 17, 1971).

Appendix 3-3 Overall Thermal Resistances of Glazings

A. Windows (vertical), no indoor shades, no storm sash			
		Thermal Resistance ($m^2 \cdot K/W$)	
Type		Winter	Summer
Single glass, clear,			
surface emittance of	0.84	0.16	0.17
	0.60	0.17	0.18
	0.40	0.19	0.20
	0.20	0.22	0.23
Insulating glass, double,			
5 mm air space, 3 mm glass		0.29	0.27
6 mm air space, 3 mm glass		0.30	0.29
13 mm air space, 6 mm glass		0.36	0.31
Insulating glass, triple,			
6 mm air space, 3 mm glass		0.45	0.40
B. Windows (vertical), no indoor shades, glass outdoor storm sash with 25 mm air space			
		Thermal Resistance ($m^2 \cdot K/W$)	
Type		Winter	Summer
Single glass, clear,			
surface emittance of	0.84	0.43	0.36
	0.60	0.37	0.29
	0.40	0.40	0.29
	0.20	0.43	0.36
Insulating glass, double,			
5 mm air space, 3 mm glass		0.48	0.43
6 mm air space, 3 mm glass		0.50	0.45
13 mm air space, 6 mm glass		0.56	0.45
Insulating glass, triple,			
6 mm air space, 3 mm glass		0.67	0.56
C. Adjustment factors for windows (multiply R-values in parts A and B by these factors)			
		Wood Frame	Metal Frame
Single glass		1.05 - 1.18	0.91 - 1.00
Double glass		1.00 - 1.11	0.77 - 0.83
Triple glass		1.00 - 1.05	0.67 - 0.77

Continued

Storm sash applied		
over single glass	1.00 – 1.11	0.71 – 0.83
Storm sash applied		
over double or triple glass	1.00 – 1.05	0.67 – 0.77

D. Greenhouse Glazings

Material (includes surface resistance)	Unit Area Thermal Resistance, $m^2 \cdot K/W$
Glass, single	0.16
Glass, double, 6 mm air space	0.25
Polyethylene film, single	0.14
Polyethylene film, double, 100 mm space	0.25
Fiber glass reinforced panel	0.15
Double panel, acrylic or polycarbonate	0.35

Notes:

a. Winter conditions are –18℃ outdoor air, 21℃ indoor air, and 24 km/h (6.67 m/s) wind speed.

b. Summer conditions are 32℃ outdoor air, 24℃ indoor air, solar radiation of 782 W/m^2, and 12 km/h (3.33 m/s) wind speed.

c. Manufacturer's data should be consulted for more accurate data for specific products, especially for thermally improved windows.

d. Date obtained from the ASHRAE Handbook of Fundamentals. Consult that reference for more detail and other glazing configurations.

Appendix 3-4 Overall Thermal Resistances of Doors

A. Wood doors, no storm doors

Type	Thermal Resistance ($m^2 \cdot K/W$) Winter	Summer
Flush doors		
hollow core, 35 mm thick	0.37	0.39
hollow core, 44 mm thick	0.38	0.40
with single glazing	0.28	0.33
solid core, 35 mm thick	0.45	0.46
solid core, 44 mm thick	0.53	0.55
with single glazing	0.38	0.40
with double glazing	0.48	0.49
solid core, 57 mm thick	0.65	0.68

Continued

with single glazing	0.43	0.44
with double glazing	0.53	0.55
Panel doors		
11 mm panels, 35 mm thick door	0.31	0.33
11 mm panels, 44 mm thick door	0.33	0.34
with single glazing	0.26	0.28
with double glazing	0.35	0.37
29 mm panels, 44 mm thick door	0.45	0.46
with single glazing	0.29	0.30
with double glazing	0.40	0.42

B. Wood doors, with storm doors, winter conditions only

Type	Thermal Resistance ($m^2 \cdot K/W$) Wood storm door	Metal storm door
Flush doors		
hollow core, 35mm thick	0.59	0.55
hollow core, 44 mm thick	0.61	0.55
with single glazing	0.53	0.49
solid core, 35 mm thick	0.68	0.63
solid core, 44 mm thick		
with single glazing	0.61	0.55
with double glazing	0.70	0.65
solid core, 57 mm thick	0.88	0.84
with single glazing	0.65	0.61
with double glazing	0.76	0.70
panel doors		
11 mm panels, 35 mm thick door	0.53	0.48
11 mm panels, 44 mm thick door	0.55	0.49
with single glazing	0.49	0.43
with double glazing	0.57	0.52
29 mm panels, 44 mm thick door	0.68	0.63
with single glazing	0.52	0.46
with double glazing	0.63	0.57

References

ACGIH. 2000. *Threshold limit values for chemical substances and physical agents and biological exposure indices.* American Conference of Governmental Industrial Hygienists: Cincinnati, OH.

Agriculture Canada. 1983. Publication 1601/E. Additional info needed.

Albright, L. D. 1990. Environmental Control for Animals and Plants, Chapter 2. American Society of Agricultural and Biological Engineers (ASABE): St. Joseph, MI. *Reprinted with permission from ASABE.*

ASABE Standards. 2007. Heat Moisture Production Data of Animals. American Society of Agricultural and Biological Engineers (ASABE): St. Joseph, MI. Pg. 602-619. *Reprinted with permission from ASABE.*

ASAE Standards. 2000. Heat Moisture Production Data of Animals. American Society of Agricultural Engineers (ASAE): St. Joseph, MI.

ASHRAE Handbook of Fundamentals. 2005. Chapter 8, Thermal Comfort, Table 4. American Society of Heating, Refrigeration and Air-Conditioning Engineers (ASHRAE): Atlanta, GA. *Reprinted with permission from ASHRAE.*

ASHRAE Handbook of Fundamentals. 2005. Chapter 10, Environmental Control for Animals and Plants, Tables 5, 6, 7, and 8. American Society of Heating, Refrigeration and Air-Conditioning Engineers (ASHRAE): Atlanta, GA. *Reprinted with permission from ASHRAE.*

ASHRAE Handbook of Applications. 2007. American Society of Heating, Refrigeration and Air-Conditioning Engineers (ASHRAE): Atlanta, GA.

Canada Plan Service. 1992. M-8710, 1985; M-9707, 1992. Ottawa, Canada.

DOE. 2008. U. S. Department of Energy Databook (http://buildingsdatabook.eere.enrgy.gov/). Prepared for the Buildings Technologies Program, Energy Efficiency and Renewable Energy, United States Department of Energy, by D&R International, Ltd.; under contract to the National Energy Technology Laboratory: Washington, DC. September.

Donham, J. K., P. Haglind, Y. Peterson, R. Rylander, and L. Belin. 1989. *Environmental and health studies of workers in Swedish swine confinement buildings.* British Journal of Industrial Medicine 40: 31-37. BMJ Publishing

Group, Ltd.: London, United Kingdom.

EIA. 2005. http://buildingsdatabook. eren. doe. gov. Energy Information Agency: Washington, DC.

EPA. 1996. *Radon: Is your family at risk?* Environmental Protection Agency: Washington, DC.

NRC. 2006. Ventilation Schemes. National Research Council: Ottawa, Ontario, Canada.

Zhang, Y. 2005. Indoor Air Quality Engineering. CRC Press: Boca Raton, FL.

Zhang, Y. 1994. Swine Building Ventilation: A Guide for Confinement Swine Housing in Cold Climates. ISBN 0-9698426-0-0. Prairie Swine Center: Saskatoon, Saskatchewan, Canada.

Author's Biography

Yuanhui Zhang is Professor and Leader of the Bioenvironmental Engineering Division, and is Associate Head of the Department of Agricultural and Biological Engineering. He is also a professor of the Departments of Mechanical Science Engineering, Civil and Environmental Engineering, and Bioengineering, at the University of Illinois at Urbana-Champaign (UIUC). He is a registered professional engineer. He is a Fellow of the American Society of Heating, Refrigeration and Air-Conditioning Engineers (ASHRAE) and is the winner of the 2008 Henry-Geise Award from the American Society of Agricultural and Biological Engineers (ASABE). His research areas include indoor air quality engineering, volumetric particle tracking velocimetry, and hydrothermal liquefaction of biomass into crude oil. He teaches "Bioenvironmental Engineering" and "Indoor Air Quality Engineering" at China Agricultural University and UIUC. He is the author of the textbook "Indoor Air Quality Engineering."

Acknowledgement

This text is a compilation of years of teaching by the author at the University of Illinois at Urbana-Champaign and China Agricultural University as a sophomore class. Over this period of time, many graduate students contributed to the homework problems, laboratory exercises and valuable comments. The author is especially grateful to Douglas Barker, a Ph. D. student and a JBT Fellow, for his comprehensive review and meticulous comment.